Microlenses

SERIES IN OPTICS AND OPTOELECTRONICS

Series Editors: **E Roy Pike**, Kings College, London, UK
Robert G W Brown, University of California, Irvine, USA

Recent titles in the series

Microlenses: Properties, Fabrication and Liquid Lenses
Hongrui Jiang and Xuefeng Zeng

Handbook of Silicon Photonics
Laurent Vivien and Lorenzo Pavesi (Eds.)

Laser-Based Measurements for Time and Frequency Domain Applications: A Handbook
Pasquale Maddaloni, Marco Bellini, and Paolo De Natale

Handbook of 3D Machine Vision: Optical Metrology and Imaging
Song Zhang (Ed.)

Handbook of Optical Dimensional Metrology
Kevin Harding (Ed.)

Biomimetics in Photonics
Olaf Karthaus (Ed.)

Optical Properties of Photonic Structures: Interplay of Order and Disorder
Mikhail F Limonov and Richard De La Rue (Eds.)

Nitride Phosphors and Solid-State Lighting
Rong-Jun Xie, Yuan Qiang Li, Naoto Hirosaki, and Hajime Yamamoto

Molded Optics: Design and Manufacture
Michael Schaub, Jim Schwiegerling, Eric Fest, R Hamilton Shepard, and Alan Symmons

An Introduction to Quantum Optics: Photon and Biphoton Physics
Yanhua Shih

Principles of Adaptive Optics, Third Edition
Robert Tyson

Optical Tweezers: Methods and Applications
Miles J Padgett, Justin Molloy, and David McGloin (Eds.)

Thin-Film Optical Filters, Fourth Edition
H Angus Macleod

Laser Induced Damage of Optical Materials
R M Wood

Microlenses
Properties, Fabrication
and Liquid Lenses

Hongrui Jiang
Xuefeng Zeng

CRC Press
Taylor & Francis Group
Boca Raton London New York

CRC Press is an imprint of the
Taylor & Francis Group, an **informa** business

A TAYLOR & FRANCIS BOOK

CRC Press
Taylor & Francis Group
6000 Broken Sound Parkway NW, Suite 300
Boca Raton, FL 33487-2742

First issued in paperback 2020

ISBN 13: 978-0-367-57649-3 (pbk)
ISBN 13: 978-1-4398-3669-9 (hbk)

Library of Congress Cataloging-in-Publication Data

Jiang, Hongrui.
    Microlenses : properties, fabrication, and liquid lenses / authors, Hongrui Jiang, Xuefeng Zeng.
        pages cm. -- (Series in optics and optoelectronics)
        Includes bibliographical references and index.
        ISBN 978-1-4398-3669-9 (hardback)
        1. Optical MEMS. 2. Liquid lenses. 3. Microfluidics. I. Title.

    TK8360.O68J53 2013
    681'.423--dc23   2012050906

**Visit the Taylor & Francis Web site at
http://www.taylorandfrancis.com**

**and the CRC Press Web site at
http://www.crcpress.com**

Dedictions

To Guangyun, Jessica, Sophia, and Haosheng.

Hongrui Jiang

To my parents.

Xuefeng Zeng

Contents

About the Authors

Hongrui Jiang is a professor in the Department of Electrical and Computer Engineering, a faculty affiliate in the Department of Biomedical Engineering, a faculty member of the Materials Science Program, and a member of the McPherson Eye Research Institute at the University of Wisconsin–Madison. He earned his bachelor's degree in physics from Peking University, China, and his master's and PhD degrees in electrical engineering from Cornell University, Ithaca, New York. His research interests include microelectro-mechanical systems, especially microscale optical systems, microfluidics, nanoscale science and engineering, smart materials and micro-/nano-struc-tures, and biomimetics and bioinspiration.

Xuefeng Zeng, PhD, is a senior engineer in Global Foundries Inc. He was a former research associate at the University of Wisconsin–Madison. He earned his bachelor's and master's degrees in microelectronics from Tianjin University and Tsinghua University, China, respectively, and his master's and PhD degrees in electrical engineering from the University of Wisconsin–Madison, Madison, WI. His research interests are in microelectromechani-cal systems, micro optics, microfabrication and defect inspection.

Preface

The history of microlenses spans more than 400 years. With the development of microscale fabrication methods, microlenses have in the past decades drawn a lot of interest and found many applications, many of which are unique. The intriguing feature of most of these microlenses is that they all involve liquids: some are formed directly from liquids; some use liquids during operation; some are solid but their fabrication involves liquids. Therefore, in this book, we aim to examine the recent progress in the emerging and fascinating field of liquid-based microlenses.

The book starts with a review of microlenses and relevant physics and fabrication methods. Then it describes various microlenses with non-tunable and tunable focal lengths based on different mechanisms. Finally, challenges surrounding current microlenses are discussed. The book is organized into eight chapters.

Chapter 1, *Introduction to Liquid Microlenses,* starts by describing the ubiquitous problems in optics; in particular, those that can be solved by liquid microlenses. To whet the appetites of readers, we point to the benefits and advantages of liquid microlenses in many applications: for example, artificial implementation of compound eyes; endoscopic fiber microscopes for confocal reflectance and real-time optical coherence tomography (OCT); photolithography; optical communications; orthoscopic imaging systems; and labs on chips.

In Chapter 2, *Basic Physics of Liquid Microlenses,* we first provide a review of the physics involved in microlenses. Beginning with light- and material-related parameters such as refractive index, absorption, and reflectance, we introduce optical lenses and their parameters (focal length, aberrations, depth of focus, and resolution). The design of lenses is also discussed. Then the chapter continues with a review of surface tension because most microlenses involve liquid–liquid or air–liquid interfaces at some point. Throughout the discussion in this chapter, the microscale associated with these liquid microlenses is emphasized.

Chapter 3, *Fabrication Methods,* introduces briefly the fabrication methods involved in microlenses. The chapter covers fabrication from substrate materials to procedures and details facility and equipment requirements for general fabrication methods. In addition to discussion of basic microfabrication procedures, namely deposition, photolithography, and etching, we also give short descriptions of other useful fabrication techniques such as lift-off, annealing, liquid-phase photopolymerization, micromolding, soft lithography, electroplating, sacrificial processes, bonding, surface modification, laser-assisted processes, planarization, and fabrication on flexible substrates and curved surfaces.

Chapter 4, *Solid Microlenses,* focuses on examples whose focal lengths cannot be tuned during their operations. Microlenses with fixed focal lengths whose focal points are shifted through displacement are also discussed in this chapter.

These non-tunable microlenses include Ge/SiO_2 core/shell nanolenses; glass lenses made by isotropic etching; self-assembled lenses and lens arrays; lenses fabricated by direct photo-induced polymerization; lenses formed by thermally reflowing photoresist; lenses formed from inkjet printing; arrays fabricated through molding processes; and injection-molded plastic lenses.

Unlike traditional glass lenses, focal lengths of microlenses can be tuned by various electrical and mechanical techniques. Chapter 5, *Electrically Driven Tunable Microlenses*, describes several examples of electrically tuned microlenses including liquid-crystal-based lenses; and liquid lenses driven by electrostatic forces, dielectrophoretic forces, electrowetting, and electrochemical reactions.

Chapter 6, *Mechanically Driven Tunable Microlenses*, continues with examples of microlenses tuned by a variety of mechanical methods. These mechanically tunable microlenses can be categorized as thin-membrane lenses with varying apertures, pressures, and surface shapes; swellable hydrogel lenses; liquid–liquid interface lenses actuated by environmentally stimuli-responsive hydrogels; and oscillating lens arrays driven by sound waves.

In these chapters, most emerging tunable and non-tunable microlenses based on various mechanisms have their optical axes perpendicular to their substrates, thus requiring optical alignments of different layers. This means that complicated structures are required for applications such as labs on chips. Chapter 7, *Horizontal Microlenses Integrated in Microfluidics*, presents examples of microlenses whose optical axes are parallel to substrates of microfluidic networks. These horizontal microlenses include two-dimensional polymer lenses; tunable and movable liquid droplets as lenses; hydrodynamically tuned cylindrical lenses; liquid core and liquid cladding lenses; air–liquid interface lenses; and tunable liquid gradient refractive index lenses.

Chapter 8, *Looking into the Future*, summarizes the importance of microlenses and attempts to shed light on future work on microlenses and related challenges such as the packaging of systems, effects of gravity, evaporation of liquids, aberrations, and integration with other optical components.

We want to express our gratitude to many people for making this book possible. The materials are obviously drawn from the works of many other researchers. We thank these colleagues for their great effort that has tremendously advanced the field of microlenses. We also wish to thank Aditi Kanhere, Dr. Ye Liu, Chenhui Li, and Xiudong Wu for helping prepare some of the figures in the book. Daniel Christensen, the laboratory manager of the Wisconsin Center for Applied Microelectronics, graciously allowed us to take photos of the clean room facility and some of the equipment described in this book. Finally, we are grateful to John Navas for his interest in this topic and motivating the writing of this book and to Rachel Holt, Luna Han, and Prudence Board for editorial support throughout the process.

Hongrui Jiang
Xuefeng Zeng

1

Introduction to Liquid Microlenses

This chapter starts by describing the ubiquitous problems in optics, in particular, those that can be solved by liquid microlenses. Then we will discuss the benefits and advantages of liquid microlenses in many applications: artificial implementation of compound eyes; endoscopic fiber microscopes for confocal reflectance and real-time optical coherence tomography (OCT); photolithography; optical communications; imaging systems; and labs on chips.

1.1 History of Microlenses

The history of microlenses with fixed focal length may be traced back to the seventeenth century. Hooke melted small filaments of glass and allowed the surface tension in the molten glass to form the smooth spherical surfaces required for lenses [1]. The principle is very simple and has now been repeated by melting other patterned materials such as photoresists and ultraviolet light (UV)-curable epoxies, and making them reflow at high temperatures to form lenses or lens arrays [2–7].

Advances in technology have enabled microlenses to be designed and fabricated by a variety of methods. Biebuyck et al. introduced a method to form microlenses, also by utilizing surface tensions [8]. Self-assembled monolayers (SAMs) are used to pattern hydrophobic and hydrophilic regions. Water and immiscible hydrocarbon fluids (oils) exhibit different wettability characteristics in these two regions. Water droplets are segregated from hydrocarbon fluid at the surface of the SAM and thus form lenses.

Pan et al. utilized melted plastics to fabricate microlenses. At the melting temperature, the hot melted plastic is intruded into circularly shaped holes and stopped at the desired depth under elevated temperature and pressure. After cooling down, the microlenses are formed. This is called a hot intrusion process [9].

Choo et al. used a different method. Microlenses are "printed" by a commercial inkjet printer. By controlling the volume of each droplet spread from a nozzle in the printer, the liquid microlenses can be printed onto most surfaces or structures [10].

Another approach involved etching silicon or glass at the same speed along different directions to form a master with hemispherical holes. Microlenses can then be formed by molding or embossing from this master [4,11]. The ability to fabricate arrays containing thousands or millions of precisely spaced lenses has led to an increased number of applications.

1.2 Categories of Microlenses

Microlenses are small; their diameters are generally about a millimeter (mm) [12] or smaller. They are categorized as single microlenses and microlens arrays containing multiple lenses in a one- or two-dimensional (1D or 2D) array. Microlenses also can be divided into fixed focal length and tunable focus types.

Based on working mechanisms, microlenses can be divided into three categories. Figure 1.1 shows the three kinds. A typical microlens has a single element with one plane surface and one spherical convex surface that can refract light. Another type of microlens has two parallel surfaces, and the focusing ability is realized by a variation in the refractive index across the lens material; these are called gradient index (GRIN) lenses. The third group, designated micro Fresnel lenses, focus light by refraction in a set of concentric surfaces with stepped edges or multiple levels [13]. The intent is to create many smaller sections of a lens with the same curvature but far thinner so that each section assumes similar focusing capability without consuming more material and thus has lighter weight. Note that Fresnel lenses are sometimes confused with Fresnel zone plates that can also serve as lenses. They are capable of focusing by diffracting light constructively and interfering at desired focus levels. The typical lens is most popular and has been studied for a long time. Because of certain advantages, GRIN and Fresnel lenses have begun to attract a great deal of interest.

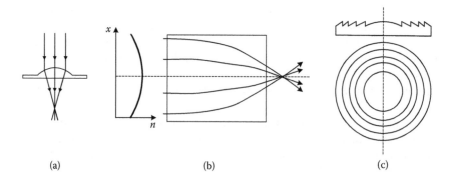

(a)   (b)   (c)

FIGURE 1.1
(a) Typical microlens. (b) GRIN lens. (c) Micro Fresnel lens.

1.3 Physics of Microlenses

A wide variety of methods for realizing microlenses with fixed or tunable focal lengths have been reported. Most of them involve air–liquid and/or immiscible liquid–liquid interfaces at the microscale. In microlenses sized around a millimeter or smaller, surface tension dominates gravity and the process of producing nearly perfect spherical shapes using a liquid–liquid interface is straightforward. More details about surface tension are introduced in Section 2.3. These interfaces are smooth and provide good optical properties [14].

1.4 Microlens Arrays

Microlens arrays contain multiple lenses formed in a 1D or 2D array on a supporting substrate. Usually individual lenses have circular apertures and are not designed to overlap. The best way to arrange them on a substrate to obtain maximum coverage of the area is to place them in a hexagonal array. However, there will still be gaps between the lenses that may be reduced only by making the microlenses with non-circular apertures [15]. Therefore, fill factor is an important characteristic determining the performance of a microlens array.

Fill factor is defined as the ratio of the active refracting area, i.e., the area that directs light to the photosensors, to the total contiguous area occupied by the complete microlens array including any gaps. To increase the effective fill factor in an optical sensor array, tiny lenses are fabricated directly onto the image sensor chip in front of each single unit cell or pixel to concentrate light onto the photosensitive region so that the light is not wasted on non-photosensitive areas [16,17].

1.5 Ubiquitous Problems in Optics

The miniaturization of optoelectronic multimedia devices led to significant reduction in sizes of both the sensor pixels and optical components so that these parts are approaching their physical limits. Diffraction effects become noticeable and decreasing light sensitivity causes image degradation due to noise as pixel size is reduced. Although these effects may be tolerable in consumer digital photography, they affect the performance of subsequent image processing and analysis techniques when used for machine vision and industrial inspection processes [18].

1.6 Applications of Liquid Microlenses

With advances in miniaturization, microlenses continue to play important roles in many fields, including artificial implementation of compound
eyes [2,19], optical communication [20,21], photolithography [3,22], biomedical imaging systems [2,23,24], orthoscopic three-dimensional (3D) imaging
[25,26], and labs on chips [4,27–29].

1.6.1 Artificial Implementation of Compound Eyes

Compound eyes in nature present intriguing subjects in physiological optics
because of their unique optical schemes for imaging. Artificial implementation of compound eyes has attracted a great deal of research interest because
the wide field of view (FOV) of a compound eye exhibits a huge potential for
medical, industrial, and military applications [2].

The use of the microlenses fabricated by using various methods has been
proposed to simultaneously mimic the structure and function of an individual ommatidium and a large-scale collection of ommatidia. Planar artificial neural compound eyes were first realized by Tunnermann's group in
Germany [18,19,30]. Compared with the natural compound eyes, the planar
eye has a smaller FOV but significantly less fabrication complexity. By using
a novel 3D polymer-based fabrication method, Lee's group successfully presented a curved compound eye consisting of a microlens array on a curvilinear surface with corresponding waveguides. Their device was comparable to
natural compound eyes in both structure and function [2].

1.6.2 Endoscopic Fiber Endoscopes for Confocal Reflectance and Real-Time Optical Coherence Tomography (OCT)

OCT is an optical imaging technique that has been coupled successfully with
endoscopy to allow scanning inside the body at depths of typically a few
millimeters [31,32]. OCT is a high-resolution, non-destructive optical technique used for cross-sectional imaging of biological tissue [33]. Most conventional endoscopic OCT probes use lenses with fixed focal lengths as the
focusing objectives.

This has a few drawbacks. First, it is difficult to adjust the focal point to the
desired position manually, particularly for irregular and inhomogeneous
tissue [34]. Second, there is an inverse relation between the lateral resolution
and the depth of focus (DOF) in OCT. Third, biological tissue is generally a
highly scattering optical medium [35]. For this reason, it is difficult to detect
the backscattered light from the tissue, particularly for structures located far
from the focal point [32].

One means to circumvent these problems is to shift the focal point con-
comitantly with the axial scan of OCT [32,36,37]. An efficient approach for

shifting the focal length is to use tunable optical components instead of moving the entire massive focusing objective [32,37–40].

If a tunable lens with variable focus is employed in the objective, many of these limitations in OCT performance may be circumvented. As the focal length is tuned with the axial scan such that the focused position always overlaps with the interference position, a number of advantages accrue: (1) the DOF may be reduced so that the beam waist is small and the reflection corresponds only to a small volume in the sample, improving interference contrast; (2) reflection from other parts of the sample is suppressed; (3) the position of maximum intensity moves with the axial scan into the sample; high lateral resolution is maintained [32].

1.6.3 Photolithography

Conventional photolithography, holographic lithography, electron beam lithography, and laser pattern writing are used commonly to fabricate repetitive micropatterns [41]. Although these techniques produce high-quality patterns, they require expensive facilities and involve multistep processing. By using microlens arrays, Wu et al. demonstrated a simple method to generate micropatterns with submicron resolution [6]. This technique includes two related but distinct methods: (1) microlens lithography using collimated illumination and (2) microlens lithography using patterned illumination [41,42].

1.6.4 Optical Communications

Free space optical interconnection networks are preferable over other methods, especially for cases of relatively high interconnection volumes. Free space systems make full use of parallelism since all three spatial dimensions are involved. However, dynamic systems are necessary for telecommunication applications, especially for optical fiber communications.

There are two distinguished kinds of dynamic free-space optical switching approaches. Optoelectronic switching architecture involves layers of elementary optoelectronic devices and requires significant power because of the data regeneration. An alternative approach consists of ensuring full data transparent switching, including two main architectures, namely the shuttered architecture and the steering architecture. The optical crossbar switch can guide light from inlet to outlet, based on shuttered geometries such as the vector–matrix or matrix–matrix multiplier. The light signal coming from each inlet is distributed by means of a programmable shutter to guide light into a corresponding outlet. Each programmable shutter can merely be a pixel of an amplitude spatial light modulator or a microlens array [20].

Moreover, steering architectures may minimize losses. They provide programmable beam steering elements that can direct the incident light into the required outlets in real time. The performance of the system depends only on the beam steering elements. Microlens arrays have been used to improve the switching systems based on dynamic optical fiber interconnection systems [20,21].

1.6.5 Orthoscopic Imaging Systems

The recording and display of true three-dimensional (3D) images was pioneered by Lippmann using 2D microlens arrays to create integral images recorded on photographic films [43]. Recently a number of potentially practical imaging systems capable of presenting true 3D optical models have been reported [25,26,44,45]. Efficient large aperture lens elements are required for 3D integral imaging and display systems. By employing an optical combination composed of macroscale lens arrays in conjunction with microlens-based focusing screens, a suitable transmission element that retains the required angular and lateral resolution can be constructed. The advantage of using segmented lenses lies in their capacity to produce spatially inverted, scaled images suitable for direct orthoscopic capture [45].

Integral photography uses an array of microlenses to encode a range of views of a 3D object. A recording is made with a photographic plate or electronic image sensor placed in the focal plane of the lens array. To replay the 3D image, the light paths are reversed, and the lens array generates light beams that intersect and form the "integral" image [26].

1.6.6 Lab-on-Chip Devices

A lab-on-a-chip (LOC) is a device that integrates one or several laboratory functions on a single chip of square millimeters to a few square centimeters in size. LOCs bear other names such as micro total analysis systems (μTASs) and microfluidics. LOC technology is undergoing rapid development and numerous promising applications have evolved in recent years.

As the first step toward a real LOC, portable devices using optical fibers have been developed using various materials and technologies. The main drawback arises from the energy losses of the excitation light due to numerous connecting parts such as the weak coupling between the external light source and the fiber or between the fiber and microfluidic channels [29].

Miniaturizing optical components such as light sources, waveguides, lens detectors, and switches, and integrating them into microfluidics is a profound approach to enhancing LOC systems, especially for biological sensing, chemical analysis, microscopic imaging, and information processing [27]. As important optical components, micro-optical lenses greatly aid further miniaturization of functionally complex LOC systems. Several current micro-optical lenses integrated within microfluidic devices are based on different mechanisms such as 2D polymer lenses, tunable and movable lenses formed through static air–liquid interfaces, dynamic optofluidic microlenses controlled by varying velocity ratios of different flows, and tunable liquid GRIN lenses. Unlike other microlenses, those integrated in microfluidic devices have a unique character in that their optical axes are perpendicular to the substrates. Therefore, their mechanism, design, and fabrication procedures are different, and we will discuss these lenses in Chapter 7.

1.6.7 Other Applications

Solid microlenses and microlens arrays have been studied and used for more than twenty years [12]. Some applications of microlens arrays such as beam shaping [46], focusing light onto CCD arrays [47], and Shack-Hartmann wavefront sensors [10] have been commercialized. Applications of solid microlens arrays are covered extensively in Daly's book [12]. In this book, we will cover some new applications and fabrications of solid microlenses and microlens arrays in Chapter 4, and will focus on tunable microlenses.

References

1. R. Hooke, *Micrographia*. London: The Royal Society of London, 1665.
2. K. H. Jeong, J. Kim, and L. P. Lee, "Biologically inspired artificial compound eyes," *Science*, vol. 312, pp. 557–561, Apr 2006.
3. M. H. Wu and G. M. Whitesides, "Fabrication of diffractive and micro-optical elements using microlens projection lithography," *Advanced Materials*, vol. 14, pp. 1502–1506, Oct 2002.
4. J. C. Roulet, R. Volkel, H. P. Herzig, E. Verpoorte, N. F. de Rooij, and R. Dandliker, "Fabrication of multilayer systems combining microfluidic and microoptical elements for fluorescence detection," *Journal of Microelectromechanical Systems*, vol. 10, pp. 482–491, Dec 2001.
5. W. Moench and H. Zappe, "Fabrication and testing of micro-lens arrays by all-liquid techniques," *Journal of Optics A: Pure and Applied Optics*, vol. 6, pp. 330–337, Apr 2004.
6. M. H. Wu and G. M. Whitesides, "Fabrication of two-dimensional arrays of microlenses and their applications in photolithography," *Journal of Micromechanics and Microengineering*, vol. 12, pp. 747–758, Nov 2002.
7. A. Jain and H. K. Xie, "An electrothermal microlens scanner with low-voltage large-vertical-displacement actuation," *IEEE Photonics Technology Letters*, vol. 17, pp. 1971–1973, Sep 2005.
8. H. A. Biebuyck and G. M. Whitesides, "Self-Organization of Organic Liquids on Patterned Self-Assembled Monolayers of Alkanethiolates on Gold," *Langmuir*, vol. 10, pp. 2790–2793, Aug 1994.
9. L.-W. Pan, X. Shen, and L. Lin, "Microplastic lens array fabricated by a hot intrusion process," *Journal of Microelectromechanical Systems*, vol. 13, pp. 1063–1071, Dec 2004.
10. H. Choo and R. S. Muller, "Addressable microlens array to impove dynamic range of Shack-Hartmann sensors," *Journal of Microelectromechanical Systems*, vol. 15, pp. 1555–1567, Dec 2006.

11. K. Wang, K.-S. Wei, M. Sinclair, and K. F. Böhringer, "Micro-optical components for a MEMS integrated display," in *The 12th International Workshop on The Physics of Semiconductor Devices (IWPSD)*, Chennai, India, 2003.

12. D. Daly, *Microlens Arrays*. Boca Raton, FL: CRC Press, 2000.

13. T. Fujita, H. Nishihara, and J. Koyama, "Micro Fresnel lenses fabricated by electron-beam lithography," *Electronics and Communications in Japan*, vol. 64-C, pp. 104–110, 1981.

14. C. Y. Yang, F. H. Ho, and J. A. Yeh, "Observation of liquid/liquid interface by atomic force microscopy," in *The 15th International Conference on Solid-State Sensors, Actuators and Microsystems* Denver, CO, USA, 2009, pp. 2050–2053.

15. C. T. Pan and C. H. Su, "Fabrication of gapless triangular micro-lens array," *Sensors and Actuators A-Physical*, vol. 134, pp. 631–640, Mar 2007.

16. Y. I. a. K. Tanigaki, "A high photosensitivity IL-CCD image sensor with mono-lithic resin lens array," in *Proceedings of IEEE Int. Electronic Device Meeting*, 1983, pp. 497–500.

17. C. L. J. N.T. Gordon, and D. R. Purdy, "Application of microlenses to infrared detector arrays," *Infrared Physics*, vol. 31, pp. 599–604, 1991.

18. A. Bruckner, J. Duparré, P. Dannberg, A. Brauer, and A. Tunnermann, "Artificial neural superposition eye," *Optics Express*, vol. 15, pp. 11922–11933, Sep 2007.

19. J. Duparré, P. Dannberg, P. Schreiber, A. Brauer, and A. Tunnermann, "Thin compound-eye camera," *Applied Optics*, vol. 44, pp. 2949–2956, May 2005.

20. H. Hamam, "A two-way optical interconnection network using a single mode fiber array," *Optics Communications*, vol. 150, pp. 270–276, May 1998.

21. S. Eitel, S. J. Fancey, H. P. Gauggel, K. H. Gulden, W. Bachtold, and M. R. Taghizadeh, "Highly uniform vertical-cavity surface-emitting lasers integrated with microlens arrays," *IEEE Photonics Technology Letters*, vol. 12, pp. 459–461, May 2000.

22. C. Y. Chang, S. Y. Yang, M. S. Wu, L. T. Jiang, and L. A. Wang, "A novel method for fabrication of plastic microlens array with aperture stops for projection photolithography," *Japanese Journal of Applied Physics Part 1-Regular Papers Brief Communications & Review Papers*, vol. 46, pp. 2932–2935, May 2007.

23. S. Kuiper and B. H. W. Hendriks, "Variable-focus liquid lens for miniature cam-eras," *Applied Physics Letters*, vol. 85, pp. 1128–1130, Aug 2004.

24. S. H. Chen, X. J. Yi, L. B. Kong, M. He, and H. C. Wang, "Monolithic integration technique for microlens arrays with infrared focal plane arrays," *Infrared Physics & Technology*, vol. 43, pp. 109–112, Apr 2002.

25. R. F. Stevens, N. Davies, and G. Milnethorpe, "Lens arrays and optical system for orthoscopic three-dimensional imaging," *Imaging Science Journal*, vol. 49, pp. 151–164, 2001.

26. R. F. Stevens and T. G. Harvey, "Lens arrays for a three-dimensional imaging system," *Journal of Optics A: Pure and Applied Optics*, vol. 4, pp. S17–S21, Jul 2002.

27. L. Dong and H. Jiang, "Tunable and movable liquid microlens in situ fabricated within microfluidic channels," *Applied Physics Letters*, vol. 91, p. 041109, Jul 2007.

28. N. Chronis, G. L. Liu, K. H. Jeong, and L. P. Lee, "Tunable liquid-filled micro-lens array integrated with microfluidic network," *Optics Express*, vol. 11, pp. 2370–2378, Sep 2003.

29. S. Camou, H. Fujita, and T. Fujii, "PDMS 2D optical lens integrated with microflu-idic channels: principle and characterization," *Lab on a Chip*, vol. 3, pp. 40–45, 2003.

30. J. Duparré, P. Dannberg, P. Schreiber, A. Brauer, and A. Tunnermann, "Artificial apposition compound eye fabricated by micro-optics technology," *Applied Optics*, vol. 43, pp. 4303–4310, Aug 2004.
31. Z. Yaqoob, J. Wu, E. J. McDowell, X. Heng, and C. Yang, "Methods and application areas of endoscopic optical coherence tomography," *Journal of Biomedical Optics*, vol. 11, p. 063001, 2006.
32. K. Aljasem, A. Werber, A. Seifert, and H. Zappe, "Fiber optic tunable probe for endoscopic optical coherence tomography," *Journal of Optics a-Pure and Applied Optics*, vol. 10, p. 044012, Apr 2008.
33. D. Huang, E. A. Swanson, C. P. Lin, J. S. Schuman, W. G. Stinson, W. Chang, M. R. Hee, T. Flotte, K. Gregory, C. A. Puliafito, and J. G. Fujimoto, "Optical Coherence Tomography," *Science*, vol. 254, pp. 1178–1181, Nov 1991.
34. N. Iftimia, B. Bouma, J. de Boer, B. Park, B. Cense, and G. Tearney, "Adaptive ranging for optical coherence tomography," *Optics Express*, vol. 12, pp. 4025–4034, 2004.
35. L. V. Wang, "Optical tomography for biomedical applications," *Engineering in Medicine and Biology Magazine, IEEE*, vol. 17, pp. 45–46, 1998.
36. F. Lexer, C. K. Hitzenberger, W. Drexler, S. Molebny, H. Sattmann, M. Sticker, and A. F. Fercher, "Dynamic coherent focus OCT with depth-independent transversal resolution," *Journal of Modern Optics*, vol. 46, pp. 541–553, 1999.
37. X. Zeng, C. T. Smith, J. C. Gould, C. P. Heise, and H. Jiang, "Fiber endoscopes utilizing liquid tunable-focus microlenses actuated through infrared light," *Journal of Microelectromechanical Systems*, vol. 20, pp. 583–593, 2011.
38. B. Qi, A. P. Himmer, L. M. Gordon, X. D. V. Yang, L. D. Dickensheets, and I. A. Vitkin, "Dynamic focus control in high-speed optical coherence tomography based on a microelectromechanical mirror," *Optics Communications*, vol. 232, pp. 123–128, Mar 2004.
39. A. Divetia, T.-H. Hsieh, J. Zhang, Z. Chen, M. Bachman, and G.-P. Li, "Dynamically focused optical coherence tomography for endoscopic applications," *Applied Physics Letters*, vol. 86, pp. 103902–103903, 2005.
40. T. Xie, S. Guo, Z. Chen, D. Mukai, and M. Brenner, "GRIN lens rod based probe for endoscopic spectral domain optical coherence tomography with fast dynamic focus tracking," *Optics Express*, vol. 14, pp. 3238–3246, 2006.
41. M. H. Wu, K. E. Paul, and G. M. Whitesides, "Patterning flood illumination with microlens arrays," *Applied Optics*, vol. 41, pp. 2575–2585, May 2002.
42. M. H. Wu and G. M. Whitesides, "Fabrication of arrays of two-dimensional micropatterns using microspheres as lenses for projection photolithography," *Applied Physics Letters*, vol. 78, pp. 2273–2275, Apr 2001.
43. G. Lippmann, "Epreuves reversible," *Photog. Integr. Comp. Rend.*, vol. 146, pp. 446–451, 1908.
44. R. F. Stevens and N. Davies, "Lens arrays and photography," *J. Photogr. Sci.*, vol. 39, pp. 199–208, 1991.
45. Neil A. Davies, M. McCormick, and M. M. a. M. Brewin, "Design and analysis of an image transfer system using microlens arrays," *Opt. Eng*, vol. 33, p. 3624–3633, 1994.
46. A. W. Snyder, D. J. Mitchell, L. Poladian, and F. Ladouceur, "Self-induced optical fibers: spatial solitary waves," *Opt. Lett.*, vol. 16, pp. 21–23, 1991.
47. T. D. Binnie, "Fast imaging microlenses," *Applied Optics*, vol. 33, pp. 1170–1175, 1994.

2

Basic Physics of Liquid Microlenses

This chapter first provides a brief review of the physics involved in microlenses, particularly, optics. Refractive index, absorption, and reflectance of materials are dependent on the wavelength of light and are discussed. The relationship between lens parameters such as resolution, field of view, focal length, aberrations, depth of focus, image quality, and designs are also discussed. The chapter continues with a review of surface tension because most microlenses involve the liquid–liquid or air–liquid interfaces at some point.

2.1 Light

2.1.1 Light Ray Tracing

Visible light covers a range of electromagnetic radiation that can be detected by the human eyes. Its wavelength ranges from about 380 to about 740 nm. The area of optics (studying light) is normally divided into three sub-areas: geometric optics, physical optics, and spectroscopy. Although we will discuss microlenses in this book, most microlenses are much larger than the wavelength of visible light and thus diffraction can be neglected except for the analysis of resolution. Thus, in this book we mainly focus on geometric optics and most simulation results are based on light ray tracing using rays to model the propagation of light through an optical system.

A good understanding of the propagation of light rays is helpful for tracing their trajectories when they pass through various optical media. As shown in Figure 2.1, lines drawn in the direction of propagation of light are called rays. However, light rays do not exist in the real world, and are only for use in theory and descriptions of how light propagates. Surfaces perpendicular to rays are called wavefronts.

The propagation of light rays through optical elements is described by a simple 2 × 2 matrix [1]. For example, propagation through a homogeneous medium with length d can be written as

$$\begin{bmatrix} r_{out} \\ r'_{out} \end{bmatrix} = \begin{bmatrix} 1 & d \\ 0 & 1 \end{bmatrix} \begin{bmatrix} r_{in} \\ r'_{in} \end{bmatrix} \tag{2.1}$$

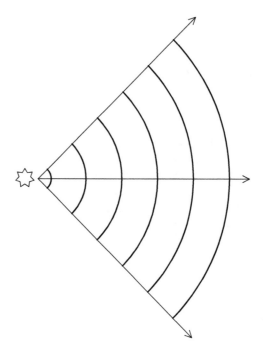

FIGURE 2.1
Spherical wavefronts emerging from a point source. Light rays are normal to wavefronts.

where r_{in} and r_{out} are the positions of the ray from the optical axis and r'_{in} and r'_{out} are the slopes of the rays relative to the optical axis. In the matrix approach, each ray is represented by a ray vector, while optical elements are represented by 2 × 2 ray matrices. Table 2.1 shows the matrices of some optical elements. The output ray vector is obtained by multiplying the ray matrices in sequence with the input ray vector.

2.1.2 Refractive Index

In optics, the refractive index (or index of refraction) of a material or medium is considered a normalized value of the speed of propagation of light in the medium. It is a ratio of the speed of light in a vacuum relative to that in the considered medium and can be written mathematically as:

$$n = c/v_p \tag{2.2}$$

where c is the speed of light in a vacuum and v_p is speed of light in the medium. Refractive index is also equal to

$$n = \sqrt{\varepsilon_r \mu_r} \tag{2.3}$$

TABLE 2.1

Ray Matrices for Typical Optical Elements and Media

Straight section with length d	$\begin{bmatrix} 1 & d \\ 0 & 1 \end{bmatrix}$
Thin lens: focal length f ($f > 0$, converging; $f < 0$, diverging)	$\begin{bmatrix} 1 & 0 \\ \frac{-1}{f} & 1 \end{bmatrix}$
Dielectric interface: refractive indices n_1 and n_2	$\begin{bmatrix} 1 & 0 \\ 0 & \frac{n_1}{n_2} \end{bmatrix}$
Spherical dielectric interface: radius R	$\begin{bmatrix} 1 & 0 \\ \frac{n_2 - n_1}{n_2 R} & \frac{n_1}{n_2} \end{bmatrix}$
Spherical mirror: radius of curvature R	$\begin{bmatrix} 1 & 0 \\ \frac{-2}{R} & 1 \end{bmatrix}$

where ε_r is the relative permittivity of the medium and μ_r is its relative permeability. For most natural materials, μ_r is very close to 1 at optical frequencies and thus n is approximately $\sqrt{\varepsilon_r}$.

Another definition of the refractive index comes from the refraction of a light ray entering a medium. The refractive index is defined as the ratio of sines of the incident angle θ_1 and the refracted angle θ_2 as light passes into the medium, as expressed in Equation (2.4).

$$n = \frac{\sin \theta_1}{\sin \theta_2} \tag{2.4}$$

The angles are measured to the normal of the surface. This definition is based on Snell's law and is equivalent to the definition above if the light enters from the reference medium. The refractive indices of materials are not constants and vary based on certain parameters such as temperature and especially the wavelength (frequency) of light, which is called dispersion. Dispersion may cause the focal lengths of lenses to be wavelength dependent. This is one source of chromatic aberration that requires correction in imaging systems. Chromatic aberration is discussed in some detail in Section 2.2.4.5.

In regions of the spectrum where the material does not absorb, the refractive index tends to decrease with increasing wavelength and thus increase with frequency. This is called normal dispersion, in contrast to anomalous dispersion in which the refractive index increases with wavelength. For visible light, normal dispersion means that the refractive index is higher for blue light than for red light. For optics in the visual light range, the amount of dispersion of a lens material is often quantified by the Abbe number:

$$V = \frac{n_{yellow} - 1}{n_{blue} - n_{red}} \qquad (2.5)$$

The relationship between the refractive indices and wavelengths of most materials can be found in handbooks. Because of dispersion, it is usually important to specify the vacuum wavelength at which a refractive index is measured.

2.1.3 Complex Refractive Index and Absorption

When light passes through a medium, some of the light is absorbed. This can be conveniently taken into account by defining a complex index of refraction:

$$\tilde{n} = n + i\kappa \qquad (2.6)$$

Here, the real part of the refractive index n indicates the phase velocity, while the imaginary part κ indicates the amount of absorption loss when the electromagnetic wave propagates through the medium material. Both n and κ are relative to the frequency. When $\kappa > 0$, light is absorbed. When $\kappa = 0$, light travels infinitely without loss. An absorption spectrum of a material can be used to determine its full complex refractive index as a function of wavelength.

2.2 Optical Lenses

A typical lens is made of transparent material and has one or two spherical surfaces. In this book, most cases refer to typical lenses.

2.2.1 Focal Length

If light is from a focal point and is incident on a convex surface, it becomes parallel through the surface, as shown in Figure 2.2a. That focus is called the primary, first, front, or object-side focal point designated F_1. Its distance to the plane tangential to the vertex is called the first or front focal length or f_1. Conversely, if parallel light is incident on the surface, it could converge toward the secondary, second, back or image-side focal point known as F_2. The corresponding distance is called the second or back focal length (f_2) as shown in Figure 2.2b. Figures 2.2c and d show the definitions of focal points and focal lengths for a concave surface.

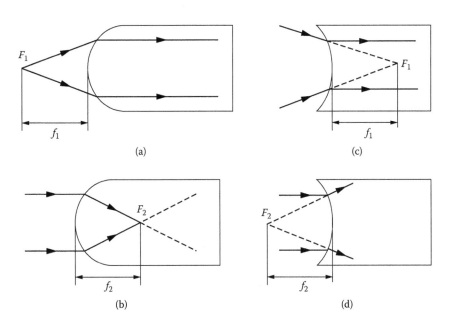

FIGURE 2.2
Focal point F and focal length f.

Planes constructed at the focal points perpendicular to the optical axis are called focal planes. Parallel rays, even if not parallel to the axis, intersect in the focal plane. Corresponding points in the object and image space—points that satisfy an object–image relation—are called conjugate points. Planes within the points are conjugate planes. The distances between conjugate points and the surface are called conjugate distances [2].

2.2.2 Image Formation

It is assumed that the medium on the left of a spherical surface has a refractive index of n_1 and the medium on the right has an index of n_2 as shown in Figure 2.3. An object is located at point O and its conjugate point is I. A light ray emerging from O with an angle of γ is refracted by the surface. For paraxial rays with small angles, the sine and tangent of an angle are approximately equal to the angle itself. Hence, we have

$$\sin\alpha = \alpha = \gamma + \phi \approx \frac{h}{o} + \frac{h}{R} \tag{2.7}$$

and

$$\sin\beta = \beta = \phi - \gamma' \approx \frac{h}{R} - \frac{h}{i} \tag{2.8}$$

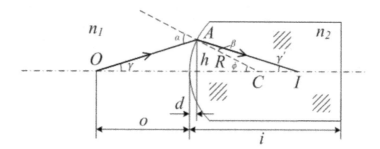

FIGURE 2.3
Refraction at single spherical surface.

Equations (2.7) and (2.8) are then substituted into Snell's law,

$$\frac{\sin \alpha}{\sin \beta} = \frac{n_2}{n_1} \qquad (2.9)$$

to obtain

$$\frac{n_1}{0} + \frac{n_2}{i} = \frac{n_2 - n_1}{R} = \frac{\Delta n}{R} \qquad (2.10)$$

which is Gauss' formula for refraction at a single refracting surface. Here Δn is the difference between the refractive index of the material of the lens and that of the outside medium, and R is the radius of curvature of the surface.

The refractive power of a surface is defined as the ratio of the refractive index of the medium over the focal length, as expressed in Equation (2.11). A surface or lens of short focal length has high refractive power, and vice versa.

$$P = \frac{n}{f} = \frac{\Delta n}{R} \qquad (2.11)$$

2.2.3 Thin Lenses

A lens is defined as thin if a ray entering on one surface emerges at approximately the same position on the opposite surface; that is, negligible translation occurs in the lens. However, there is no numerical limit distinguishing thin and thick lenses. The issue is decided entirely by the degree of precision required for solving a given problem. One lens may be considered thin for a preliminary and thick for a rigorous solution. A thin lens can be considered the sum of two single refractive surfaces:

$$P = P_1 + P_2 = \frac{\Delta n}{R_1} + \frac{\Delta n}{R_2} \qquad (2.12)$$

$$P = \frac{1}{f} = \Delta n \left(\frac{1}{R_1} + \frac{1}{R_2} \right) \qquad (2.13)$$

Their powers added together represent the lens maker's formula. Note that in Equation (2.13), n is assumed to be 1.

2.2.4 Aberrations

2.2.4.1 Spherical Aberration

Spherical aberration arises when rays passing through different zones of a lens come to different focal points. In general, rays close to the optic axis are refracted less and come to a focus further away from the lens than the marginal rays, as shown in Figure 2.4. This is called positive longitudinal spherical aberration, and a lens acting this way is called under-corrected. If the marginal rays have a longer focal length, the spherical aberration is negative and the lens is over-corrected.

A microlens can be formed through a liquid–liquid interface. The longitudinal spherical aberration is the difference between the focal length calculated from Equation (2.13) and the intersection point of the real light rays (O'). The transverse spherical aberration is the projection of the longitudinal spherical aberration on the focal plane.

Here we present a simple model to describe spherical aberrations of microlenses [3]. As an example, the geometry of a microlens is shown in Figure 2.5. The microlens is formed through the interface between deionized water and silicone oil. The refractive indices of the water and oil are $n_2 = 1.33$ and $n_1 = 1.48$, respectively. The diameter of the microlens d is a constant. The radius of curvature of the water–oil interface R varies. O is the center of the interface

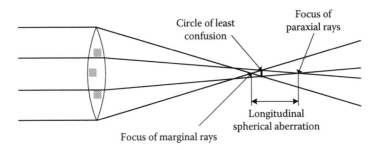

FIGURE 2.4
Positive longitudinal and lateral spherical aberration.

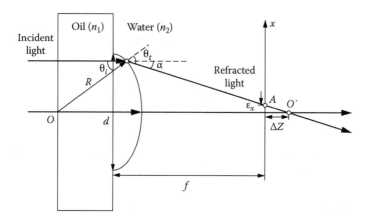

FIGURE 2.5
Physical geometry of microlens. (*Source:* Zeng, X.F. et al. 2010. *Journal of Micromechanics and Microengineering*, 20, 115035. With permission from Institute of Physics.)

curvature. θ_i, θ_t, and α are the incident angle, refracted angle, and difference between the two angles, respectively. The focal length of the microlens f equals

$$f = \frac{R}{n_1 - n_2} \tag{2.14}$$

and the x axis is in the focal plane of the microlens. Two refracted light rays from the center and edge of the microlens intersect at the point of O'. The distance between the intersection point O' and the focal plane is ΔZ, the longitudinal spherical aberration [4]. The light ray from the edge of the microlens strikes the focal plane at point A. The distance from point A to the optical axis OO' is ε_x, the transverse spherical aberration [4].

$$\varepsilon_x = \Delta Z \cdot \tan \alpha = \frac{d}{2} - \left(f + R(1 - \cos \theta_i) \right) \cdot \tan \alpha \tag{2.15}$$

where

$$\alpha = \theta_t - \theta_i \tag{2.16}$$

and

$$n_1 \cdot \sin \theta_i = n_2 \cdot \sin \theta_t \tag{2.17}$$

2.2.4.2 Coma Image

The term *coma* is derived from the comet-like appearance of the image of a point object that is located off the axis. In spherical aberration, various zones of a lens produce a longitudinal difference of focus for rays parallel to the

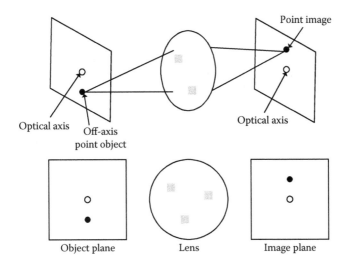

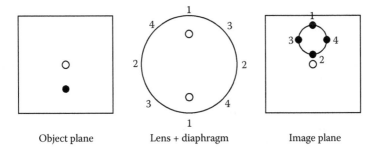

FIGURE 2.6
Coma schematic.

FIGURE 2.7
Coma and comatic circle.

lens axis. If the incident rays have an angle with the axis, coma will appear. Therefore, even if a lens system has been corrected for spherical aberration, images of points off the axis will not be sharp, as shown in Figure 2.6.

Imagine that an opaque diaphragm with two holes numbered 1-1 is placed in front of the lens. Light passing through the two holes 1-1 will come to a focus (#1 on the screen), as shown in Figure 2.7. If the diaphragm with the two holes is then rotated 90 degrees to position 2-2, the light will come to a focus at #2. For all four positions (1-2-3-4) of the diaphragm, the light will be focused into a 1-2-3-4 (comatic) circle rather than into a single point. In Figure 2.7, we can notice how the image points 1-2-3-4 are distributed around the comatic circle in relation to the orientation of the holes in the diaphragm.

Larger zones in the lens through which the light passes can result in larger comatic circles, as shown in Figure 2.8. In fact, the radius of a given comatic

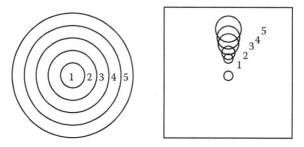

FIGURE 2.8
Larger zones producing larger comatic circles.

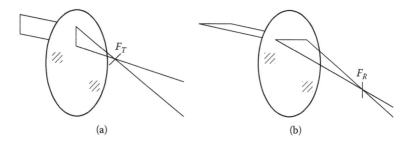

FIGURE 2.9
Bands of light move to different focal points, depending on their orientation. (a) Tangential focus. (b) Radial focus.

circle is proportional to the square of the radius of the zone related to it. The distance from the center of the comatic circle to the optic axis is proportional to the square of the radius of the zone. The figure with all comatic circles produced by light passing through the whole lens is called a comatic flare, as shown in Figure 2.8. The flare may point toward or away from the axis, depending on the type of lens producing the aberration.

2.2.4.3 Astigmatism

When a narrow bundle of light goes through a spherical surface some distance away from the optic axis, the light will form an ellipse on the surface, with the major axis pointing toward the vertex of the surface and the minor axis at a right angle to it. Rays lying in the major axis will come to a focus at point F_T, called the tangential focus, as shown in Figure 2.9a. Rays in the minor axis come to a focus at F_R, the radial focus, as shown in Figure 2.9b. Thus, if an off-axis point object is imaged through a simple surface or lens, two focal lines will result: the radial line that lies in the plane of incidence and the tangential line perpendicular to the plane. This separation is called astigmatism.

If the object is a radial line, its image is sharp at F_R and blurred at F_T. If the line object is rotated about the axis and a series of successive images are

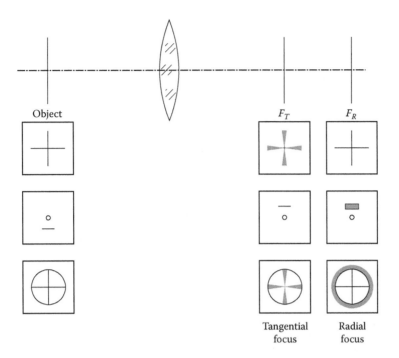

FIGURE 2.10
Radial line object of rotational symmetry (top), tangential line object (center), and astigmatic image of spoked wheel (bottom).

recorded, we could obtain the same result, but with rotational symmetry, as shown in Figure 2.10. If the off-axis object is a tangential line, it comes to a focus at F_T, not at F_R. Thus, if a complex object is like a spoke wheel, with the axis of the wheel coincident with the optic axis, the rim of the wheel comes to a focus at F_T and the spokes at F_R, as shown in Figure 2.10.

2.2.4.4 Distortion

Distortion, similar to coma, is not a primary aberration but results from other aberrations. In distortion, the transverse linear magnification in the image varies with the distance from the optical axis, which results in distortion and causes a square object to look like a barrel or a pincushion, as shown in Figure 2.11. Distortion cannot change the resolution of the image, but it changes the locations of image points. For a sufficiently thin lens there is no distortion.

2.2.4.5 Chromatic Aberration

The index of refraction of a transparent medium varies with wavelength, as discussed in Section 2.1.2. Therefore, a single lens has different focal lengths for different colors of light. With a simple positive lens, blue light comes to a

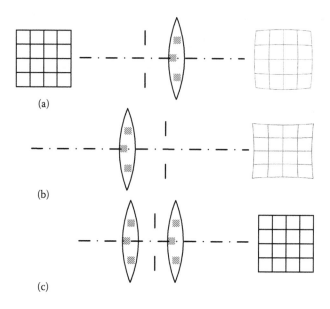

FIGURE 2.11
(a) Barrel distortion. (b) Pincushion distortion. (c) Symmetric doublet with central stop is relatively free from distortion. Effects are reversed with a negative lens.

focus closer to the lens than red light. This horizontal distance between the axial images is called axial or longitudinal chromatic aberration. The vertical difference in the height of the images resulting from the axial difference is called lateral chromatic aberration. An image seen in the "blue focus" will show blue details surrounded by a red halo. In the "red focus," the details are red and the halo is blue. Chromatic aberration is called positive if the blue focus lies closer to the lens. With a diverging lens, the effect is reversed.

The relationship between wavelength and focal length is shown in Figure 2.12. The magnitude of chromatic aberration can be expressed, like dispersion, as the difference between the focal length for red light (C line) and that for blue light (F line).

2.2.4.6 Measurement of Aberrations

As discussed in the previous section, aberrations are important characteristics of lenses and affect the overall performance of the optical system integrating the lenses. There are a few ways to measure the aberrations. First, if the geometric profiles of the lens and the materials of which the lens is made (refractive indices) are known, ray tracing can predict how the rays would be bent as they pass through the lens and thus calculate the aberrations. This essentially utilizes the definitions of the aberrations described above. To obtain a lens profile, mechanical profilometers [5,6] or white light interferometry [5,7] can be used.

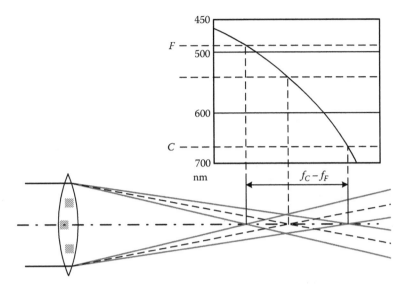

FIGURE 2.12
Chromatic aberration and plot of focal length versus wavelength.

The second approach to measuring aberrations is by interferometry. The aberrations are, after all, essentially imperfections that appear when the lens focuses the incoming light and causes distortion in the output wavefront. Such deformation would be reflected from the interference patterns between the output wave and a reference wave (usually a plane wave). The measured interference patterns can be used in turn to determine the aberrations, usually using Zernike polynomials [8–10].

Many types of interferometry can be used for this purpose. For instance, the Twyman-Green interferometer [11] is suitable for surface testing in reflection. The Mach-Zehnder interferometer permits lens testing in transmission [12]. The third interferometric technique utilizes a Shack-Hartmann wavefront sensor [10,13,14]. This method can be applied to both reflection and transmission. Interestingly, the Shack-Hartmann sensor utilizes a microlens array in front of a CCD imager array to create a series of focal spots that are recorded by the CCD imager to obtain the wavefront.

Most of the above methods to measure the aberrations of lenses are suitable for solid lenses. Special care must be taken, though, in the design of a system to characterize microlenses, especially the diffraction patterns owing to the small dimensions of the lenses (discussed later) that are superimposed with interference fringes [11]. Some of the methods that apply to liquid microlenses that are formed via liquid–air or liquid–liquid interfaces are difficult to apply or the setup must be carefully redesigned to adjust to the liquid nature. For instance, using a mechanical profilometer to obtain a lens profile becomes impractical. A common technique for reflection-based interferometry that involves depositing a highly reflective layer onto the lens surface also becomes impractical.

2.2.5 Depth of Focus and Resolution

Depth of focus is the range of distances between object points imaged with acceptable sharpness and the image plane. Depth of focus has two slightly different definitions. The first definition is the distance in which the image plane can be displaced while a single object plane remains in acceptably sharp focus. The second one is the image-side conjugate of depth of field. Depth of focus in the first definition is symmetrical on the image plane, while in the second definition it is greater on the far side of the image plane; in most cases, the distances are approximately equal.

Depth of focus increases with smaller apertures. For distant subjects (beyond macro range), depth of focus is relatively insensitive to focal length and subject distance for a fixed f number (defined as the ratio of the focal length of the lens to the diameter of its entrance pupil). In the macro region, depth of focus increases with longer focal length or closer subject distance, while depth of field decreases.

We wish to add a few more words on the f number, which is an important value for a lens. It is often denoted as F/#, and bears other names such as focal ratio, f ratio, f stop, or relative aperture. The f number of a lens can be easily changed even though its focal length is fixed. For example, the entrance pupil size varies when irises of different openings are placed before the lens. Real lenses cannot focus all light rays perfectly. Even at best focus, a point is imaged as a spot rather than a point. A circle of confusion is an optical spot resulting from a cone of light rays from a lens that does not achieve a perfect focus when imaging a point source. Circle of confusion can be used to determine depth of field, the part of an image that is acceptably sharp.

When the depth of focus relates to a single plane in object space, it can be calculated

$$t = 2Nc\frac{v}{f} \tag{2.18}$$

where t is the total depth of focus, N is the lens f number, c is the circle of confusion, v is the image distance, and f is the lens focal length. In most cases, the image distance is not easily determined and thus the depth of focus can also be given in terms of magnification m,

$$t = 2Nc(1 + m) \tag{2.19}$$

The magnification depends on the focal length and the subject distance. When the magnification is small, the formula can simplify to

$$t \approx 2Nc \tag{2.20}$$

The circle of confusion is taken as the lens focal length divided by 1000 [15]. This formula is valid only for normal lenses. The depth of focus Δf is the amount of defocus that introduces a wavefront error, and can be calculated using

$$\Delta f = \pm 2\lambda N^2 \tag{2.21}$$

Resolution depends on the distance between two distinguishable radiating points. Here we assume an adequate level of contrast in a mathematical model of Airy discs. However, real optical systems are complex and it is difficult to increase the distance between distinguishable point sources.

The resolution of a system is based on the minimum distance γ at which the points can be distinguished as individuals. On the line between the center of one point and the next, the contrast between the maximum and minimum intensity should be at least 26% lower than the maximum. This corresponds to the overlap of one Airy disk on the first dark ring in the other. This standard for separation is also known as the Rayleigh criterion. The distance is expressed as:

$$\gamma = \frac{1.22\lambda}{2n\sin\theta} = \frac{0.61\lambda}{NA} \tag{2.22}$$

where γ is the minimum distance between resolvable points, λ is the wavelength of light, n is the index of refraction of the media surrounding the radiating points, θ is the half angle of the pencil of light that enters the objective, and NA is the numerical aperture.

The lens resolution is usually determined by the quality of the lens but is limited ultimately by diffraction. Light coming from a point in the object diffracts through a lens aperture and forms a diffraction pattern in the image with a central spot and surrounding bright rings, separated by dark nulls. This pattern is called the Airy pattern, and the central bright lobe is the Airy disk. The angular radius of the Airy disk is given by

$$\sin\theta = 1.22\frac{\lambda}{D} \tag{2.23}$$

where θ is the angular resolution, λ is the wavelength of light, and D is the diameter of the lens aperture. Two adjacent points in the object produce two diffraction patterns. If the angular separation of the two points is significantly less than the Airy disk angular radius, the two points cannot be resolved in the image. On the contrary, if their angular separation is much greater than that, images of the two points are different and can therefore be resolved.

The Rayleigh criterion is considered when two points have an angular separation equal to the Airy disk radius to first null. Thus the greater the diameter of the lens or its aperture, the greater the resolution. On the other hand, for microscale lenses whose apertures are smaller, the diffraction limit significantly lowers the resolution.

2.2.6 Calculating Optical Parameters

2.2.6.1 Introduction to Zemax

Zemax is a widely used commercial optical design program developed by Zemax Development Corporation (Bellevue, Washington, U.S.). It is used for the design and analysis of optical systems. Zemax can perform standard sequential ray tracing through optical elements, non-sequential ray tracing for analysis of stray light, and physical optics beam propagation [16,17]. Zemax can be used for the design of optical systems such as camera lenses and analysis of illumination systems.

The aberrations in the previous sections can be analyzed by using Zemax. The program can model the propagation of rays through optical elements such as lenses, mirrors, and diffractive elements. Zemax can model the effects of optical coatings on the surfaces of components and produce standard analysis illustrations such as spot diagrams and ray-fan plots. The physical optics propagation feature can be used for problems in which diffraction is important—propagation of laser beams, holography, and the coupling of light into single-mode optical fibers. Zemax has a powerful suite of optimization tools that can be used to optimize a lens design by automatically adjusting parameters to maximize performance and reduce aberrations.

2.2.6.2 Spherical Aberrations of Polydimethylsiloxane (PDMS) Microlenses

As an example, we will use Zemax to calculate the spherical aberrations of microlenses made of PDMS [16]. The two materials involved are PDMS (Figure 2.13) and air. Air is the default material and its properties do not need to be changed in Zemax. The refractive index of PDMS is added into the glass catalog in Zemax. Here we choose the simple Schott formula to express the refractive index of PDMS because it is widely used by glass manufacturers. Equation (2.24) shows the expression of the Schott formula.

$$n^2 = a_0 + a_1\lambda^2 + a_2\lambda^{-2} + a_3\lambda^{-4} + a_4\lambda^{-6} + a_5\lambda^{-8} \tag{2.24}$$

The required coefficients $a_0 \dots a_5$ are available in glass catalogs. However, the refractive index of PDMS depends on the composition ratio of the mixture and the curing conditions and varies among different references [18–20]. Here we assume that the refractive index of PDMS is independent of the wavelength and is 1.416 at a wavelength of 610 nm [19]. Thus, $a_0 = 2.005$ and other coefficients equal zero.

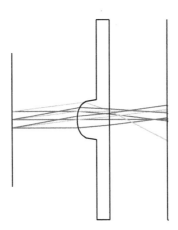

FIGURE 2.13
Layout of typical PDMS microlens with diameter of 1.2 mm. (*Source:* Zeng, X. 2009. Microlenses and Their Application in Endoscopes. Department of Electrical and Computer Engineering, University of Wisconsin, Madison. With permission.)

The monochromatic light is assumed in the simulation and the wavelength is 637 nm. There are two surfaces in each microlens. The flat surface is set to "Infinity." The surface geometry of the curvature is extracted from the side profile obtained for the microlens, and the fourth order polynomials are fitted. The surface type is set to "Even Asphere" and the fourth order polynomial coefficients of the aspherical surface are used. Figure 2.13 shows a typical diagram of the microlens. The diameter is 1.2 mm and the number of rays is 3. After setting up "Even Asphere," we proceed to "Analysis," select "Aberration Coefficients," and open the "Seidel Coefficients" window. The corresponding surface of W040 in the "Seidel Aberration Coefficients in Waves" section is the spherical aberration. For this microlens, the spherical aberration is 1.354 µm.

2.2.6.3 Liquid Microlens Formed by Water–Oil Interfacial Meniscus

As a second example, we calculate the spherical aberration of a liquid microlens formed by a water–oil interfacial meniscus [16]. Water and oil are often used for lens materials. The refractive indices of water and oil are 1.33 and 1.48, respectively and are assumed for simplicity to be independent of wavelength. The a_0 coefficients in the Schott formula for the corresponding materials are 1.769 and 2.190, respectively. The monochromatic light is used in the calculation and the wavelength is set to 637 nm.

The microlens has three surfaces. The two flat surfaces are set to "Infinity." The curved surface is in the middle and formed by the water–oil interfacial meniscus. The geometry of the curvature of this interface is extracted from the side profile and the fourth order polynomials are fitted. The same procedures are used to define this surface. Figure 2.14 shows a typical liquid

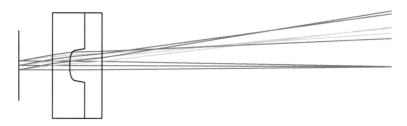

FIGURE 2.14
Layout of typical liquid microlens formed by water–oil interfacial meniscus. (*Source:* Zeng, X. 2009. Microlenses and Their Application in Endoscopes. Department of Electrical and Computer Engineering, University of Wisconsin, Madison. With permission.)

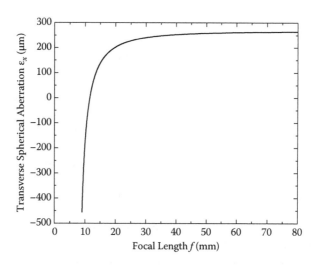

FIGURE 2.15
Relationship between transverse spherical aberration ε_x and focal length f of microlens. The calculated transverse spherical aberration equaled 0 when the focal length was 11.5 mm. (*Source:* Zeng, X.F. et al. 2010. *Journal of Micromechanics and Microengineering*, 20, 115035. With permission from Institute of Physics.)

microlens diagram. The diameter is 1.8 mm. The number of rays is 3. For a typical liquid microlens shown in Figure 2.14, the spherical aberration is −0.245 μm.

2.2.6.4 Measurements of Spherical Aberration

The transverse spherical aberration of a microlens was calculated from Equation (2.15) to Equation (2.17) as plotted in Figure 2.15. The diameter of the microlens was a constant 2.4 mm. The focal length f varied from 8.9 to 80 mm. The corresponding incident angle θ_i calculated from the focal length and microlens aperture varied from 63 to 5.5 degrees. More information about this lens will be discussed in Section 6.5. The calculated transverse spherical aberration ε_x ranged from −450 to 260 μm and equaled zero when the focal length was 11.5 mm.

Collimated light from a blue laser at a different wavelength of 473 nm illuminated a microlens from the bottom. A camera was mounted on the translation stage and moved along the optical axis of the microlens to the focal point. The camera resolution was 640 × 480 pixels and the size of each pixel was 8.3 × 8.3 μm². The focused spot images at three focal lengths and the laser spot at the distance of 25 mm were recorded by the CCD in the camera, as shown in Figure 2.16. The laser spot was formed through the microlens structure without any lens liquids. The normalized light intensities were extracted from the images and plotted as spots in Figure 2.16 [3].

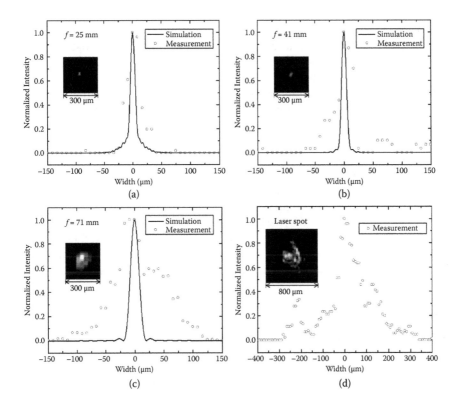

FIGURE 2.16
Normalized light intensities of focused spots of typical microlens at focal lengths of (a) 25 mm; (b) 41 mm; and (c) 71 mm. The circle curves were extracted from images taken by a camera at the focal point. The simulation results using Zemax light-ray tracing software are the black solid curves. (d) Normalized light intensity of laser spot at distance of 25 mm without lens liquids. (*Source:* Zeng, X.F. et al. 2010. *Journal of Micromechanics and Microengineering*, 20, 115035. With permission from Institute of Physics.)

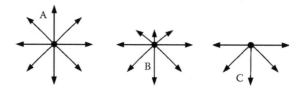

FIGURE 2.17
Liquid molecules exhibiting cohesive forces from neighboring molecules. A is attracted equally in every direction, resulting in a net force of 0. B and C are subject to net forces pulling them inward toward the bulk of the liquid.

The light intensity profiles at three microlens focal lengths were simulated by Zemax. Point spread function (PSF) analysis was performed in Zemax for the microlens, as shown as the black solid line in Figure 2.16 [3]. The distribution of light intensities at three microlens focal lengths extracted from the images matched well with the Zemax simulation (a through c in the figure). Figure 2.16d shows the light intensity of the laser spot at the distance of 25 mm without the lens. Because of the diffraction of the polymer material, the curve in d is not smooth.

Normally the spherical aberrations from normalized light intensities are defined as the full width at half maximum (FWHM). In order to compare with the calculated transverse spherical aberrations, however, we defined the spherical aberrations from the normalized light intensities as the width at minimum. The spherical aberrations at focal lengths of 25, 41, and 71 mm were 58, 166, and 290 μm, respectively. The spherical aberrations from the normalized light intensities had a good match with the result calculated from Equation (2.15) to Equation (2.17).

Before we move to the discussion on surface tension, we wish to point out that our discussion on the optics of the lenses is brief. Interested readers are referred to the literature [1,2,4] for more information and detailed analyses.

2.3 Surface Tension

Surface tension is a property of the surface of a liquid that allows it to resist an external force caused by the cohesion of similar molecules. It is responsible for many behaviors of liquids. Surface tension has a unit of force per unit length. The Système International (SI) unit is the newton per meter (N/m); the commonly used centimeter–gram–second (cgs) unit is the dyne per centimeter (dyn/cm).

Liquid molecules experience cohesive forces from their neighboring molecules (Figure 2.17). Water at 20°C has a surface tension of 72.8 dyn/cm. It is also sometimes useful to consider the surface tension in terms of work (or energy) per unit area. This is especially true for studies of the thermodynamics of a system. The SI unit in this situation is the joule per meter squared (J/m^2); the cgs unit is the erg per centimeter squared (erg/cm^2).

In a bulk liquid, the molecules (A) are attracted equally in every direction by neighboring liquid molecules, resulting in a net force of zero. Such attractive forces between like liquid molecules are often called *cohesive forces*; they can be viewed as residual electrostatic forces and called *van der Waals forces*. The molecules close to the surface (B) or at the surface (C) are not surrounded on all sides by other molecules and therefore are pulled inward (we have neglected the force exerted from the molecules in the air onto the liquid molecules as their force is much smaller).

The inward pull creates some internal pressure and forces liquid surfaces to contract to minimal area. Surface tension is responsible for the shapes of liquid droplets. Although easily deformed, droplets of water tend to be pulled into spherical shapes by the cohesive forces of the surface layer.

A molecule in contact with a neighbor is in a lower state of energy than if it were alone (not in contact with a neighbor). The interior molecules have as many neighbors as they can possibly have, but the boundary molecules lack neighbors (compared to interior molecules) and therefore have higher energy. For a liquid to minimize its energy state, the number of higher energy boundary molecules must be minimized. The lower number of boundary molecules results in a minimized surface area.

As a result of surface area minimization, a surface will assume the smoothest shape it can. Since any curvature in the surface shape results in greater area, a higher energy will also result. Consequently the surface will push back against any curvature in much the same way as a ball pushed uphill will roll back to minimize its gravitational potential energy.

For a liquid surface, if no force acts normal to a tensioned surface, the surface must remain flat. However, if the pressure on one side of the surface differs from pressure on the other side, the pressure difference times surface area results in a normal force. The surface tension forces must cancel the force due to pressure and the surface must be curved. Figure 2.18 shows how the curvature of a tiny patch of surface leads to a net component of surface tension forces acting normal to the center of the patch. When all the forces are balanced, the result is known as the Young-Laplace equation:

$$\Delta p = \gamma \left(\frac{1}{R_x} + \frac{1}{R_y} \right) \tag{2.25}$$

where Δp is the pressure difference, γ is the surface tension, and R_x and R_y are radii of curvature in the axes that are parallel to the surface.

We have discussed the cohesive forces between liquid molecules. When the attractive forces are between different molecules, e.g., between a liquid molecule and a solid molecule, they are called *adhesive* forces. For example,

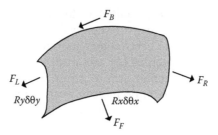

FIGURE 2.18
Surface tension forces acting on a tiny (differential) patch of surface. $\delta\theta x$ and $\delta\theta y$ indicate the amounts of bend over the dimensions of the patch.

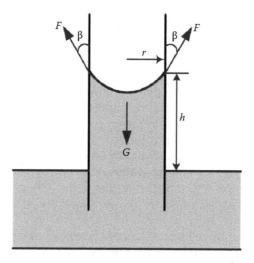

FIGURE 2.19
Capillary action. When a glass tube is inserted into water, a water column is formed in the tube. The meniscus at the tip of the water column bears the shape of a crescent.

the adhesive forces between water molecules and glass surfaces are stronger than the cohesive forces between the water molecules. As a result, water tends to be pulled along the glass surfaces. This leads to the well-known phenomenon of capillary action. As shown in Figure 2.19, a glass tube is inserted into water. The adhesive forces between water molecules and glass molecules pulls water upward until balanced by gravity. A water column is thus formed. Owing to the surface tension, the water-to-air interface is actually a curved meniscus that minimizes the surface energy. The height of the water column h can be derived to be

$$h = \frac{2F\cos\beta}{\rho g r} \qquad (2.26)$$

where F is the net force pulling the water meniscus, β is the angle between the glass surface and the tangent of the water meniscus at the contact between water and glass, ρ is the density of water, r is the radius of the glass tube, and g is the acceleration of gravity.

If the adhesive forces between the liquid and solid molecules are less than the cohesive forces between the liquid molecules, the solid surface will tend to push the liquid away. For example, if mercury replaces water in the glass tube in Figure 2.19, the mercury-to-air interface would curve upward and the mercury in the glass tube would be lower than that outside ($h < 0$).

2.3.1 Contact Angles

The contact angle is the angle at which a liquid and vapor interface at a solid surface. The equilibrium contact angle is specific for any given system and is determined by the molecular interactions across the liquid–vapor, solid–vapor, and solid–liquid interfaces. The shape of the droplet is determined by the Young-Laplace equation, with the contact angle (surface tension) defining boundary condition. Contact angle can be measured using a contact angle goniometer [21].

Figure 2.20 shows the images of a small water droplet resting on a flat horizontal polydimethylsiloxane (PDMS) surface but with surface modifications. Note that the contact angles are quite different, corresponding to the different surface properties. In Chapter 3, we will further discuss surface modification. The contact angle is not limited to a liquid–vapor interface; it is also applicable to the interfaces of two liquids.

Now let us further consider the simple example of a water droplet on a flat surface as shown in Figure 2.20. We have redrawn the boundary of the droplet in Figure 2.21. To calculate the contact angle, we note that the total force at the boundary between the droplet and the solid surface should add up to zero:

$$\gamma_{sl} - \gamma_{sv} + \gamma_{lv}\cos\theta_0 = 0 \tag{2.27}$$

where s denotes the solid phase, v the vapor phase, l the liquid phase, and γ the surface tension. The subscripts denote the interfaces; for example, γ_{lv} is the interfacial surface tension between the liquid and vapor phases. This is the Young equation. From Equation (2.27), the contact angle θ_0 can be found as

$$\cos\theta_0 = \left(\gamma_{sv} - \gamma_{sl}\right)/\gamma_{lv} \tag{2.28}$$

In the image in Figure 2.21, θ_0 is an acute angle when resting on the substrate (sometimes referred to as "mostly wetting"). If $\theta_0 > 90$ degrees instead, it is "mostly non-wetting." In general, when a liquid drop forms a spherical cap

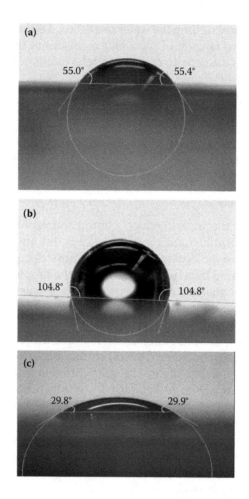

FIGURE 2.20
Images taken from goniometer examination of a water droplet on (a) native PDMS surface, (b) PDMS surface coated with octadecyltrichlorosilane, and (c) PDMS surface treated with oxygen plasma. Measured contact angles are shown.

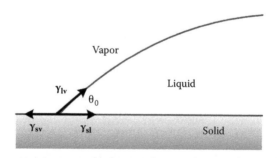

FIGURE 2.21
Boundary of liquid droplet on solid surface. At equilibrium, the net force at the boundary must be 0, leading to Young equation.

on a solid surface at equilibrium, the result is "partial wetting." The reverse is "total wetting" in which the liquid completely spreads to reach the minimum surface energy. For example, solvents such as ethanol and toluene will spread on clean glass surfaces. A solid surface that tends to have high affinity (energy) for water is often called hydrophilic. Water will spread on hydrophilic surfaces. On the other hand, if water only partially wets on a solid surface and forms a spherical cap on it with a contact angle, the surface is said to be hydrophobic because it tends to repel the water.

In practice, the *hydrophobic* and *hydrophilic* terms are often used relatively rather than quantitatively. If water has a smaller contact angle (is less repelled) on surface A than on surface B, we say that surface A is *more hydrophilic* than surface B, or surface B is *more hydrophobic* than surface A although technically both surfaces are actually hydrophobic. Another example is treating a surface so that the water contact angle increases or decreases. In this case, the treated surface is said to more hydrophobic or more hydrophilic than it was in its innate state. In some literature, the relative terms are removed. For example, one may claim that a surface is originally hydrophobic, but after a certain chemical treatment it becomes hydrophilic. Usually this means that the contact angle is reduced significantly (e.g., from 100 to 70 degrees), rather than implying that water can totally wet and spread on the new surface.

As noted previously, surface tension can also be considered as work exerted onto the surface of a liquid. Therefore, another way of writing Equation (2.27) is calculating the work done onto a water droplet by the surface tension if the line of contact moved an infinitesimal distance dx. As a result, the droplet would sweep through an infinitesimal area of dA on the solid surface. Referring to Figure 2.22, we calculate the total work done onto the droplet as

$$dW = \gamma_{sl}dA - \gamma_{sv}dA + \gamma_{lv}dA\cos\theta_0 \qquad (2.29)$$

At equilibrium, dW/dA must be zero, which leads to the same calculation as Equation (2.27).

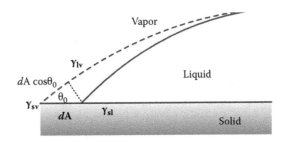

FIGURE 2.22
Another way of deriving Young equation. At equilibrium, the total work done on the liquid droplet if the line of contact moved an infinitesimal distance must be 0.

2.3.2 Contact Angle Hysteresis

Equation (2.28) suggests that the contact angle of a water droplet on a solid surface is a constant determined only by the surface tensions that are in turn dependent only on the material properties involved. This is counterintuitive. In fact, it is contrary to our experience from daily life!

Consider an experiment described in Figure 2.23a in which we first place a water droplet onto a flat solid surface. Based on the discussion above, the droplet would assume a shape and the contact angle at the boundary would be θ_0. If we tilt the solid surface, we would not expect that the droplet would keep its shape and maintain θ_0 at every point on its boundary. Rather, under the effect of gravity, the droplet should assume the shape depicted in Figure 2.23b. The lower part of the droplet has a larger contact angle θ_A than θ_0, and the opposite side of the droplet has a smaller contact angle θ_R than θ_0. As we tilt the solid surface more, θ_A should further increase and θ_R should further decrease, until the tilt reaches a critical point at which the surface tension can no longer hold the droplet and it runs down the solid surface. Figure 2.23 shows the advancing and receding contact angles, θ_A and θ_R, respectively. The apparent inconsistency between Equation (2.28) and the experiment described in Figure 2.23 arises because we previously ignored other forces; for example we ignored gravity exerted onto the water droplet in Equation (2.27). Similarly, in Equation (2.29), gravity should have been added since it would do work onto the droplet if the line of contact moved. Because the surface tension at each small section on the boundary would be in different direction as compared to gravity, we would expect the contact angle to vary too.

We could also generalize that if the droplet tends to advance at a certain point of the solid surface, the resulting advancing contact angle θ_A would be larger. If the droplet tends to recede at another point of the solid surface, the resultant receding contact angle θ_R would be smaller [22,23]. The static or thermodynamic contact angle θ_0 obtained previously would fall between the maximum θ_A and minimum θ_R.

Next we will consider another example in Figure 2.24. Two glass slides were placed opposite each other with a small gap h between them. We assume h is much smaller than the widths of the glass slides. The two surfaces of the glass

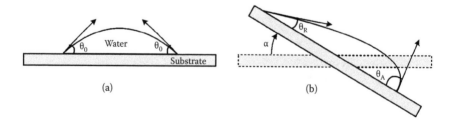

(a)   (b)

FIGURE 2.23
Advancing and receding contact angles. When a substrate is tilted, a water droplet on it will show different contact angles at different points on the boundary.

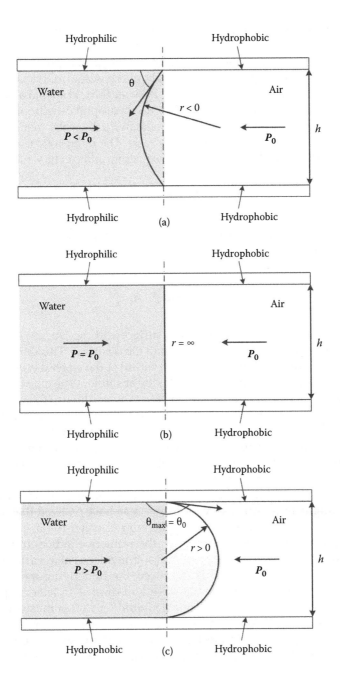

FIGURE 2.24
Controlling pressure difference across a water-to-air interface may vary the radius of curvature of the water meniscus from a negative to a positive value. (*Source:* Adapted from Cheng, D.M., Y.J.P. Choe, and H.R. Jiang. 2008. *Journal of Microelectromechanical Systems*, 17(4), 962–973. With permission.)

slides facing each other were pretreated chemically so that the left half is hydro-philic and the right half is hydrophobic with a thermodynamic contact angle of $\theta_0 > 90$ degrees. Next, we squeeze water into the gap via pneumatic pressure P from the left. The air pressure to the right of the water front is kept constant at atmospheric pressure P_0. As noted earlier, water would spread along the left half of the glass surface until it reaches the hydrophilic–hydrophobic (H-H) boundary. If initially we maintain a P less than P_0, the water meniscus would curve into the water, as shown in Figure 2.24a. The radius of curvature of the meniscus is described by Young-Laplace equation, and can be written as:

$$\Delta P = P - P_0 = \frac{\gamma}{r} \tag{2.30}$$

where γ is the surface tension. Note that initially $r < 0$, indicating that the meniscus is indeed curved into the water. From geometry,

$$r = \frac{h}{2\cos(\pi - \theta)} \tag{2.31}$$

where θ is the angle between the glass surface and the tangent of the water meniscus at the contact between water and glass surface. The above equation implies that if we vary P, thus ΔP, the curvature of the water meniscus would vary accordingly. As P is increased to P_0, the absolute value of r will increase.

Specifically, $P = P_0$ would result in a flat water-to-air interface (Figure 2.24b). If P is further increased, r becomes positive and decreases from infinity, meaning that the meniscus now protrudes further into the air. In this case, θ is an obtuse angle and increases with ΔP until it reaches θ_0 (Figure 2.24c). Further increases in P will cause the water to break the H-H boundary and burst into the hydrophobic side.

The fact that the contact angle θ in the above example can vary in a certain range is enlightening! An observant reader may have realized that a tunable lens could be formed in this way, as the curved water-to-air interface can serve as a lens due to the difference in the refractive indices of water and air; the change in the curvature of the water meniscus also varies the focal length of the resultant lens. This is indeed the underlying mechanism of many tunable liquid microlenses that we will discuss in later chapters. As we will see then, the design of a tunable liquid lens often means pinning a liquid-to-gas or liquid-to-liquid interface at a lens aperture and varying its curvature through a certain mechanism of actuation. Readers interested in the theory of surface tension are directed to a comprehensive review [25].

2.3.3 Effect of Gravity

Besides showing the concept of advancing and receding contact angles, Figure 2.23 also demonstrates that gravity plays a critical role in determining the shape of a liquid droplet. If we want to form liquid-based lenses, we must consider the effect of gravity on the shape of the lens because any distortion in its shape would affect its optical properties.

Note that the droplet shown in Figure 2.23a looks like a lens but the droplet in Figure 2.23b certainly does not. Other issues related to gravity are its omnipresence and fixed direction. These factors make it difficult to compensate for gravity in lens design because all potential orientations of the lens in operation must be considered. The best approach, therefore, is to try to minimize the effect of gravity on lens shape.

Our prior discussion on surface tension again offers some insight. Since the surface tension and adhesive forces scale down with the lengths of the liquid, solid, and vapor boundaries, they drop much slower in comparison to the gravity force on a liquid droplet that is proportional to the volume of the droplet. Hence we could envision that at some smaller scale, the surface tension would be a much more dominant force compared to gravity, and if so, the effect of gravity onto the shape of the droplet would be negligible.

Figure 2.25 shows the simulation result of the effect of gravity on the shapes of water droplets of different sizes on a silver surface in steady state (thermodynamic contact angle $\theta_0 = 90$ degrees). The simulation was performed using Surface Evolver software [26]. Figure 2.25a depicts the shape of a water droplet with a volume of 1 µL and considers the effect of gravity. Figure 2.25b depicts the shape of the same droplet when the effect of gravity is not taken into account. Similarly, Figure 2.25c shows the shape of an 8-ml water droplet, considering the effect of gravity, while Figure 2.25d shows the shape of the same droplet when the effect is ignored.

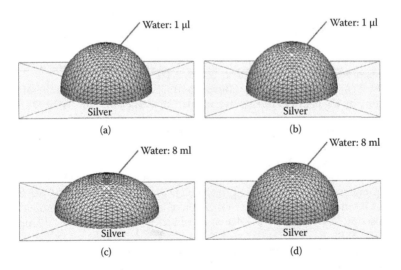

FIGURE 2.25
Simulated shapes of water droplets of different volumes on silver surfaces.

While there is no apparent difference in the shapes of the water droplet in Figures 2.25a and b, the water droplet in Figure 2.25c has sagged visibly compared to Figure 2.25d. Thus we may conclude that the effect of gravity on the shape of a droplet would be negligible compared to surface tension when the volume of a droplet approaches the order of microliters or dimensions of millimeter scale. We shall see in later chapters that the most microlenses indeed fall into this scale.

References

1. P. Y. Amnon Yariv, *Photonics: Optical Electronics in Modern Communications*. New York: Oxford University Press, 2007.
2. J. R. Meyer-Arendt, *Introduction to Classical and Modern Optics*. London: Prentice-Hall, 1972.
3. X. Zeng, C. Li, D. Zhu, H. J. Cho, and H. Jiang, "Tunable microlens arrays actuated by various thermo-responsive hydrogel structures," *Journal of Micromechanics and Microengineering*, vol. 20, p. 115035, Nov 2010.
4. R. Guenther, *Modern Optics*. Cambridge: John Wiley & Sons Inc., 1990.
5. P. Nussbaum, R. Volke, H. P. Herzig, M. Eisner, and S. Haselbeck, "Design, fabrication and testing of microlens arrays for sensors and microsystems," *Pure and Applied Optics*, vol. 6, pp. 617–636, Nov 1997.
6. G. Beadie, M. L. Sandrock, M. J. Wiggins, R. S. Lepkowicz, J. S. Shirk, M. Ponting, Y. Yang, T. Kazmierczak, A. Hiltner, and E. Baer, "Tunable polymer lens," *Optics Express*, vol. 16, pp. 11847–11857, 2008.
7. N. Chronis, G. L. Liu, K. H. Jeong, and L. P. Lee, "Tunable liquid-filled microlens array integrated with microfluidic network," *Optics Express*, vol. 11, pp. 2370–2378, Sep 2003.
8. N. Seong, G. Yeo, H. Cho, and J. Moon, "Measurement of lens aberration by using in-situ interferometer and classification of lens for correct application," *Proceedings of SPIE*, vol. 4000, pp. 30–39, 2000.
9. C. J. Evans, R. E. Parks, P. J. Sullivan, and J. S. Taylor, "Visualization of surface figure by the use of Zernike polynomials," *Applied Optics*, vol. 34, pp. 7815–7819, 1995.
10. P. D. Pulaski, J. P. Roller, D. R. Neal, and K. Ratte, "Measuremenst of aberrations in microlenses using a Shack-Hartmann wavefront sensor," *Proceedings of SPIE*, vol. 4767, pp. 1–9, 2002.
11. S. Reichelt and H. Zappe, "Combined Twyman-Green and Mach-Zehnder interferometer for microlens testing," *Applied Optics*, vol. 44, pp. 5786–5792, 2005.
12. D. Lee and R. Haynes, "Characterization of lenslet arrays for astronomical spectroscopy," *Publications of the Astronomical Society of the Pacific*, vol. 113, pp. 1406–1419, 2001.
13. T. M. Jeong, M. Menon, and G. Yoon, "Measurement of wave-front aberration in soft contact lenses by use of a Shack-Hartmann wave-front-sensor," *Applied Optics*, vol. 44, pp. 4523–4527, 2005.

14. C. Li, G. Hall, D. Zhu, H. Li, K. W. Eliceiri, and H. Jiang, "Three-dimensional surface profile measurement of microlenses using the Shack-Hartmann wavefront sensor," *Journal of Microelectromechanical Systems,* vol. 21, pp. 530–540, 2012.

15. L. Larmore, *Introduction to Photographic Principles,* 2nd ed. New York: Dover Publications, 1965.

16. X. Zeng, *Microlenses and Their Application in Endoscopes.* Department of Electrical and Computer Engineering Ph.D. Dissertation, Madison, WI: University of Wisconsin–Madison, 2009.

17. Zemax Software. http://www.radiantzemax.com/en/zemax/.

18. D. Losic, J. G. Mitchell, R. Lal, and N. H. Voelcker, „Rapid fabrication of micro- and nanoscale patterns by replica molding from diatom biosilica," *Advanced Functional Materials,* vol. 17, pp. 2439–2446, Sep 24 2007.

19. T. K. Shih, C. F. Chen, J. R. Ho, and F. T. Chuang, "Fabrication of PDMS (polydimethylsiloxane) microlens and diffuser using replica molding," *Microelectronic Engineering,* vol. 83, pp. 2499–2503, Nov-Dec 2006.

20. S. Camou, H. Fujita, and T. Fujii, "PDMS 2D optical lens integrated with microfluidic channels: principle and characterization," *Lab on a Chip,* vol. 3, pp. 40–45, 2003.

21. X. Zeng and H. Jiang, "Polydimethylsiloxane microlens arrays fabricated through liquid-phase photopolymerization and molding," *Journal of Microelectromechanical Systems,* vol. 17, pp. 1210–1217, Oct 2008.

22. C. N. C. Lam, R. Wu, D. Li, M. L. Hair, and A. W. Neumann, "Study of the advancing and receding contact angles: liquid sorption as a cause of contact angle hysteresis," *Advaces in Colloid and Interface Science,* vol. 96, pp. 169–191, 2002.

23. R. Tadmor, "Line energy and the relation between advancing, receding, and Young contact angles," *Langmuir,* vol. 20, pp. 7659–7664, 2004.

24. D. Cheng, Y. J. P. Choe, and H. Jiang, "Controlled liquid-air interfaces and interfacial polymer micromembranes in microfluidic channels," *Journal of Microelectromechanical Systems,* vol. 17, pp. 962–973, Aug 2008.

25. G. Navascués, "Liquid surfaces: theory of surface tension," *Rep. Prog. Phys.,* vol. 42, pp. 1131–1186, 1979.

26. K. Brakke, Surface Evolver Software, Version 2.50. http://www.susqu.edu/brakke/evolver/evolver.html.

3

Fabrication Methods

Before we discuss various liquid-based microlenses, it is beneficial to provide a chapter that gives a brief introduction of the methods involved in their fabrication. The fabrication of these lenses takes advantage of advancements in modern microelectronic fabrication for integrated circuits (ICs), microelectromechanical systems (MEMS), and microfluidics that enable large scale, low cost, precise manufacturing.

We also need to put fabrication into the perspectives of complete systems. Microlenses alone are of little use until they are combined with other optical components and electronics into systems. Hence, in this chapter, our descriptions of fabrication cover the basic techniques of microfabrication. However, this chapter is by no means a comprehensive discussion and readers wishing to learn more about microfabrication are encouraged to peruse other books that provide more complete reviews [1–11].

3.1 Introduction to Microfabrication Methods

Microfabrication is a process used to generate physical devices onto substrates. These devices are formed by structures with dimensions from millimeter to nanometer range. Figure 3.1 shows a piece of silicon (Si) wafer with devices after the completion of the fabrication. Over the years, microfabrication has advanced significantly from the established semiconductor fabrication processes used for integrated circuits (ICs) to diverse materials and processes such as polymers, liquids, soft lithography, and liquid-based processes.

Our discussion will start with substrates and materials used in microfabrication. Next, we present the basic approaches to defining patterns onto the substrates that ultimately lead to functional devices. These basic approaches include thin film growth and deposition, photolithography, and etching. We then briefly describe some useful special fabrication modules and techniques including lift off, annealing, liquid-phase photopolymerization, micromolding, soft lithography, electroplating, sacrificial processes, bonding, surface modification, laser-assisted processes, and planarization.

Conventional microfabrication has been based on *planar* technologies and thus limited to flat, rigid, solid surfaces. However, recent advances in technologies now allow fabrication on flexible surfaces and we will also provide examples of these devices.

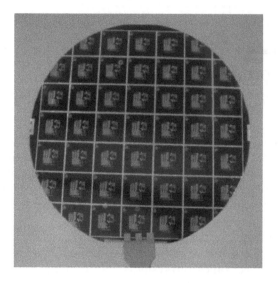

FIGURE 3.1
Silicon wafer on which devices have been fabricated.

3.2 Facilities and Equipment

Microfabrication is typically performed in a facility called a *clean room*. A clean room has low levels of environmental contaminants such as dusts, aerosol particles, and chemical vapors that can affect fabrication materials. The controlled level of contamination (class) of a clean room is specified by the number of particles per cubic meter at a specified particle size. Other environmental parameters such as temperature and humidity are also well controlled. Figure 3.2 shows a clean room.

The clean room houses many pieces of equipment used in microfabrication. Depending on the fabrication steps, contamination level tolerances vary. Therefore, different pieces of equipment are located in different areas of a clean room based on class. Figure 3.3 shows representative instruments that are typically utilized in microfabrication. Their functions will be explained in later sections of this chapter.

3.3 Substrate Materials

All microfabrication starts with a certain substrate. In some cases, the substrate merely provides a platform on which the devices will be built by layering. The layering constitutes an *additive* process. In some cases, part of the substrate is

FIGURE 3.2
Clean room facility used for microfabrication.

removed as a step of fabrication. This process is referred to as *subtractive*. In most fabrication, both additive and subtractive processes are involved.

The most common substrates used in the semiconductor industry are Si and other compound semiconductor wafers such as gallium arsenide (GaAs). These wafers are produced in various sizes and cut along different crystalline planes. They are also doped with many types and concentrations of impurities that determine whether they are n- or p-type semiconductors.

The crystalline structures and dopants of these semiconductor wafers are obviously critical factors in making semiconductor devices; however, they may not be as relevant to the fabrication of microlenses. Nevertheless, semi-conductor substrates are still important and useful for two reasons. First, most fabrication utilizes instruments developed for fabrication on these semi-conductor wafers. Second, from a system integration view, microelectronics and optoelectronics made on semiconductor wafers will be needed at some point, whether for direct integration with the lenses or subsequent assembly.

The main drawback of the semiconductor substrates is that they are not transparent in the range of visible light. This often limits aspects of the design and formation of lenses and other optical components because light transmission through substrates is not possible.

Compared to Si and other semiconductor compound substrates, glass is transparent and thus serves as a good substrate candidate for fabrication of lenses and other optical components. Common glass substrates are wafers

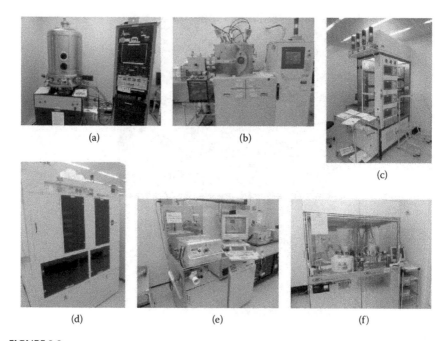

FIGURE 3.3
Equipment for microfabrication. (a) Electron beam evaporator. (b) Sputterer. (c) Low pressure chemical vapor deposition furnace. (d) Stepper for photolithography. (e) Reactive ion etching. (f) Wafer bonder.

and slides. Some techniques and instruments have been developed to process glass substrates, although the choices are more limited than techniques for processing Si wafers. Another attractive feature is that glass is biocompatible and widely used in biological studies. Therefore, making optical devices directly onto glass can expand applications for biological studies.

Other materials, especially some polymers, may also be used as the substrates. These polymers generally absorb in the infrared and ultraviolet regions but are transmissible for visible light. One example of a polymer film used as a substrate is polycarbonate (PC) plastic. It is quite rigid compared to most soft polymers but compliant and flexible compared to Si. If flexibility of a substrate is required, other polymers such as the polydimethylsiloxane (PDMS) silicone rubber can be considered. In many situations, polymer films are first formed onto glass or Si substrates and then peeled off to serve as new substrates for subsequent device fabrication.

3.4 Materials

Microfabrication utilizes a wide range of materials; thus it is impossible to prepare an exhaustive list. It is safe to state though that any material that possesses a special useful property and can be patterned into a certain microscale structure will find an application in microdevices.

For high electric and thermal conductivity, metals such as gold (Au), copper (Cu), and aluminum (Al) are used widely. Magnetic metals such as nickel (Ni) and iron (Fe) are utilized to form magnetic actuators. Some metal thin films such as chromium (Cr) and titanium (Ti) are applied to enhance the adhesion of other metal thin films to a substrate. Doped polycrystalline silicon and metal silicides [12] have electric conductivities slightly inferior to metals but much better than insulators. They have also become integral materials for microelectronics.

Dielectric inorganic materials also find widespread utilization in micro-fabrication as insulators. Common examples are silicon dioxide (SiO_2) and silicon nitride (SiN). They can also serve as masking materials during the etching steps, which we will discuss later.

Many other types of polymers contribute to microfabrication. Photoresists (PRs) are used in photolithography to define device patterns and will be described later. Many polymer materials can also be applied to form device structures, for example PDMS and poly-isobornyl acrylate (poly-IBA). A third example is SU-8, a thick PR that can form microstructures after the lithography step.

Polymer materials can be utilized to form actuators that "drive" devices to move. Good examples are stimuli-responsive hydrogels that undergo large volumetric changes (swelling and deswelling) when exposed to specific stimuli. For example, N-isopropylacrylamide (NIPAAm) responds to temperature in an aqueous environment; it swells when the temperature decreases and contracts when temperature increases. As another example, incorporating Au nanoparticles into an NIPAAm hydrogel produces an infrared (IR) light-responsive hydrogel. The Au nanoparticles can convert the photon energy upon IR irradiation to heat that in turn causes the thermal response of the NIPAAm hydrogel. Polymers are especially useful in fabricating devices whose operations involve liquids.

3.5 Basic Fabrication Steps

In this section we will discuss three basic steps in microfabrication: thin film deposition, photolithography, and etching. The whole fabrication process usually involves iterations of these steps so the device structure is built layer by layer until it is completed.

3.5.1 Thin Film Deposition

The first step to build the devices is depositing materials onto the substrates. This is usually performed by *deposition* of thin films.

The first method is *physical vapor deposition* (PVD) divided into subcategories of *evaporation* and *sputtering*. Evaporation is used primarily for deposition of metals. A metal sample is held in a container (crucible) and heated by a resistive coil made of a refractory metal, an inductive coil, or an electron (e-) beam. The metal sample is heated and evaporated as a result. A flux of the metal vapor thus reaches the substrate, where it cools and deposits a thin film of the metal onto the substrate.

Sputtering, on the other hand, relies on bombardment of ions of chemically inert atoms such as argon onto the surface of a target. These ions are formed by ionization in a plasma, followed by acceleration by an electric field. These energetic ions "knock out" atoms from the target. The ions gain enough energy to reach the substrate to deposit a thin film of the target material. Sputtering is not limited to metals; it can be applied to silicon oxide, amorphous silicon, zinc oxide, and aluminum nitride. Sputtering also offers higher deposition rate and better coverage across the topography of a substrate surface than evaporation. However, evaporation with e-beams is much more directional.

The second method is *chemical vapor deposition* (CVD). As suggested by the name, unlike PVD, chemical reactions are involved in CVD. Precursor materials in gas phases are introduced into heated furnaces and react at the substrate surface to deposit the desired thin film. For example, CVD is typically performed in low pressure conditions (< 1 Torr); this technique is called LPCVD and usually requires an inert diluent gas such as nitrogen. CVD processes typically involve high temperatures (above 500°C). This is a very important factor to consider in a designing a fabrication process. For example, no metal except tungsten (W) is allowed into CVD furnaces. LPCVD usually has very slow deposition rate. Plasma-enhanced CVD (PECVD) can deposit dielectric films much faster. It also allows deposition at lower temperatures (<400°C). This is very useful when a substrate has already been metalized.

A final remark concerns the oxidation of silicon by reacting oxygen with silicon or polycrystalline silicon to form SiO_2—also performed in a furnace. Although technically the SiO_2 is directly thermally "grown" from the Si, this process is similar to LPCVD in many aspects.

3.5.2 Photolithography

After a thin film is deposited onto a substrate, the next step is constructing a structure from this layer of thin film according to a designed pattern that must be transferred onto the substrate. This is done through a technique called *photolithography*.

The first step of photolithography is called *pattern generation*. The designed structure of a layer of thin film is imaged onto a *photomask*. The patterns on the photomask essentially define where light can pass through and where it cannot. Photomasks are typically made of various types of fused silica that have high degrees of optical transparency. The opaque layer on one side of the photomask is usually chromium.

FIGURE 3.4
Photomask with transparent and opaque regions that define device patterns.

Figure 3.4 shows a piece of a photomask. Note that the optical transparency and opacity refer to a specific wavelength; this will be discussed in more detail later. Other properties of a photomask include small thermal expansion coefficient, flatness, and highly polished surfaces to reduce scattering of light. These properties are critical for high quality and highly reproducible patterning. However, for proof-of-concept studies, other photomasks such as printed transparencies are also used.

To transfer the image of the photomask patterns to the substrate, a thin polymeric film is applied onto the substrate through a step called *spin coating*, in which drops of the material are spread over the surface of the substrate by spinning. This thin film is photosensitive and is called a *photoresist* (PR). Then the substrate undergoes a masked exposure using the photomask under light at a certain wavelength. Figure 3.5 illustrates this step of masked exposure.

This exposure step is performed using either a contact aligner (1× indicates a 1:1 ratio between the image on the photomask and the image onto the substrate) or a stepper (5× or 10×, indicating that the dimension of the image onto the substrate is reduced by a factor of 5 or 10, respectively, from the dimension of the photomask). The exposing light is usually ultraviolet (UV).

Referring to the description above, the transparency and opacity of the photomask and the photosensitivity of the PR all refer to the specific wavelength of UV light used for exposure. The PR on the substrate under the transparent regions absorbs the photon energy from the UV light and becomes soluble in a solution (developer) that is usually alkaline. Conversely, the PR masked by the photomask and not exposed to UV light remains insoluble in the developer. Therefore, after immersing the substrate into the developer, the exposed regions of the PR are removed, and the patterns from the photomask are transferred onto the substrate.

Note that the discussion here is in the context of *positive* PR. An opposite or *negative* PR is also possible; the PR becomes insoluble in the developer after exposure and remains soluble in the absence of exposure.

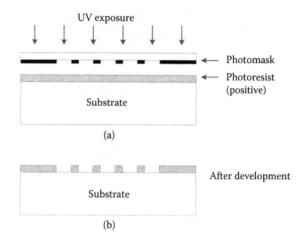

FIGURE 3.5

Photolithography process using contact aligner. (a) Ultraviolet exposure of photoresist with photomask. (b) Pattern on photomask is transferred to photoresist after development.

In our discussion, the exposure light is UV. The concept can be extended to shorter wavelengths including deep ultraviolet (DUV) light, extreme ultraviolet (EUV) light, x-rays, and even e-beams defined with de Broglie wavelength. The basic rule is that the shorter the wavelength, the finer the structure that can be defined. Complete lithographic systems, however, are very different from one another.

3.5.3 Etching

After photolithography, the designed structure of a layer is transferred onto the substrate via a PR film. However, the structure is yet to be defined from the deposited thin film. To this end, part of the thin film corresponding to the patterns of the design and the PR must be removed. This removal is called *etching*. Since we already have the PR layer bearing the designed pattern, the etching will be masked by the PR. In other words, the thin film will be removed only where there is an opening without protection from a PR.

The two main categories of etching are *wet etch* and *dry etch*. Wet etch is performed by immersing the substrate into a certain liquid chemical containing one or more suitable etchants that attack and remove the exposed thin film material. Dry etch, as the name suggests, occurs in a dry state and uses chemically reactive vapors (vapor etching) or reactive species from the gaseous phase in a plasma (plasma-assisted etching) as the etchants. One example of vapor etching utilizes xenon difluoride (XeF_2) gas that etches Si but leaves many other materials such as SiO_2 and metals intact.

Plasma-assisted etching is far more popular. The ionized species are produced in plasma, accelerated in an electric field, and directed onto the substrate surface. The etching is achieved by sputtering effects (removing material

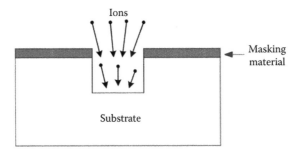

FIGURE 3.6
Reactive ion etching.

by ion bombardment) and chemical reactions. If the removal relies solely relies on sputtering, typically with an argon plasma, the process is called *ion milling*. Sometimes the ions are focused by electrodes so that they attack only a small region on the substrate. This is known as *focused ion beam milling*. When the ions have chemical reactivities that contribute to etching, the process becomes *reactive ion etching* (RIE). For example, oxygen plasma can be used to remove a PR in a process called *ashing*. Figure 3.6 shows the concept of RIE.

To this point, we assumed that the masking material for etching is a PR. This assumption is not necessary. To pattern thin film A by microfabrication, it is very common to deposit another film B on top of it and use that as the masking material for the etching of thin film A (see Figure 3.7). Of course, photolithography is done first on thin film B, and it must be subsequently etched and patterned with the PR as the masking material.

Depending on the process, the PR can be removed before the etching of thin film A, leaving thin film B as the only masking material, or the PR can be kept to serve as the masking material for thin film A along with thin film B. After etching of thin film A, the PR, and in most cases the remaining thin film B, are stripped to make way for the next structural layer. This step is blank (not masked) etching.

One important etching factor is whether the etching rate differentiates among different directions. If the etching rate is independent of the orientation of the substrate, the etching is called *isotropic* (Figure 3.8). The opposite etching technique is *anisotropic*. Most wet etching and vapor etching are isotropic. However, some etchants show crystalline orientation-dependent reactions when etching a single crystal, resulting in an anisotropic etch.

A famous example of anisotropic wet etching is etching of single crystal Si using potassium hydroxide (KOH). KOH etches Si along the {100} and {110} planes hundreds of times faster than along the {111} planes. Figure 3.9 illustrates masked etching of a single crystal Si substrate whose surface corresponds to a {100} plane. The exposed {100} plane is etched rapidly while the {111} plane is etched slowly. As the etching proceeds, two slanted sidewalls corresponding to two {111} planes appear (Figure 3.9a). If the etching

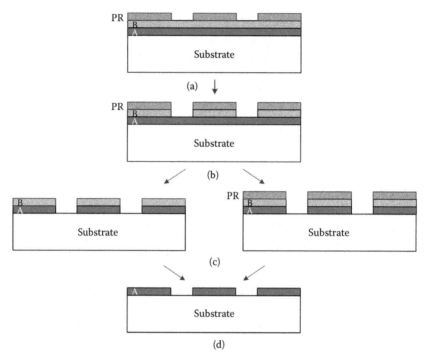

FIGURE 3.7
A thin film other than a photoresist can be used as a masking material for etching. (a) To pattern thin film A, film B is deposited on top of it. Photolithography is carried out on thin film B. (b) Etching of thin film B with photoresist as masking material. (c) Thin film B alone (left) and combined with photoeresist (right) is used as the masking material for the etching of thin film A. (d) Masking materials (thin film B, photoresist, or both) are removed.

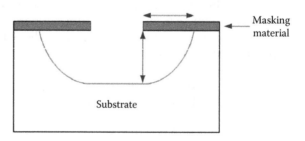

FIGURE 3.8
Isotropic etching. The etching rate is the same in all directions.

continues long enough, the two {111} planes ultimately meet, forming a V-shaped groove (Figure 3.9b). From that point, the etching and thus the expansion of the groove occur more slowly.

Dry etching is mostly anisotropic, due to the directionality of the ions. Nevertheless, in some cases, process parameters can be adjusted to realize isotropic RIE. One example is the RIE of Si with sulfur hexafluoride (SF_6).

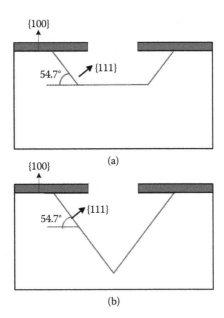

FIGURE 3.9
Anisotropic wet etching of single crystal Si using KOH. The surface of the wafer is a {100} plane.

Other important factors to consider in etching are *etch rate* and *selectivity*. Etch rate describes how fast the etching is achieved. Selectivity is the ratio between the etch rate on the target material and the rate on another material that is present. Ideally, only the target material should be etched while other materials are left intact; that is, the ideal selectivity should be infinity. Unfortunately, this is not possible in reality.

While wet etching and vapor etching usually show very high selectivities, plasma-assisted etching generally has much lower selectivity (3 to 1 would be considered very good). The finite selectivity imposes strict constraints on the design of the etching process. Care must be taken to select masking materials (the PR and/or the intermediates such as thin film B in Figure 3.7), their thicknesses, and etch lengths. For example, because of the topography of the substrate (especially after patterning a few layers), the thickness of a deposited thin film across the substrate will not be uniform. Therefore, over-etch is a common practice to ensure the removal of the thickest part of the target material. As a result, the thickness of the masking material may be well above the nominal number calculated from the nominal thickness of the target material and the selectivity.

Note that substances considered "target" and "other" materials may vary even among the sub-steps within a single etching step. Refer again to Figure 3.7. Although the ultimate target material to be etched is thin film A, it should be viewed as an "other" material when thin film B is etched and considered the "target" material.

Note also that "other" materials are not limited to masking components. For example, the layer immediately below the target material may be exposed to the etchants at some places while other thicker parts of the target material are still being etched. The selectivity between the target material and the layer immediately below must also be considered in designing an etching process. Otherwise, we could be surprised when the lower layer at some spots was etched through after the targeted etch!

Plasma-assisted etching generally has a much lower etch rate than wet etching. A special type of RIE known as *deep reactive ion etching* (DRIE) is worth mentioning. This etching proceeds in alternating steps of strong and fast RIE and polymer deposition to protect the sidewall. As a result, deep etch into the substrate is possible. For example, a Si wafer can be etched through in a few hours. Process parameters can be adjusted to obtain different etching profiles. This process is very important for the creation of high aspect ratio structures (depth >> width and/or length). This process can also be applied to glass.

3.6 Other Microfabrication Techniques

In addition to the three basic microfabrication steps discussed above, other fabrication techniques are quite useful. We will briefly discuss a few of them, namely lift-off, annealing, liquid phase photopolymerization, micromolding, soft lithography, electroplating, sacrificial processes, bonding, surface modification, laser-assisted processes, planarization, and fabrication on flexible substrates and curved surfaces. Some of these techniques do not necessarily belong to the traditional repertoire of microfabrication of ICs. However, they have proved very useful for the creation of other types of microdevices and systems such as MEMS, microfluidics, and labs on chips.

3.6.1 Lift-Off

As discussed above, the patterning of a thin film layer usually involves deposition of the thin film followed by photolithography and masked etching. However, this may not be the most convenient method for some materials, especially some metals, considering factors such as etch type (wet or dry), availability of etchants, etch rates, masking materials, and selectivity. Lift-off is an alternative method to pattern such materials while avoiding the etching step.

Figure 3.10 shows the lift-off process. A photolithography step is first performed using relatively thick PRs. The key of this photolithography step is to create a slightly re-entrant profile of the PR, meaning the top of the opening is narrower than the bottom (Figure 3.10a). This may be done by hardening

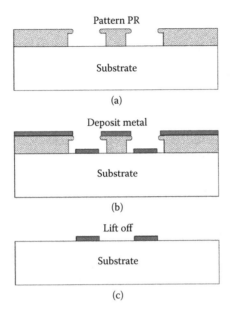

FIGURE 3.10
Lift-off process to define pattern without etching.

the PR or by using multiple layers of PRs and overdeveloping the bottom layers of the PR with the top PR layers as a mask. Next, the target thin film is deposited across the whole substrate (Figure 3.10b).

In most cases, evaporation is the chosen deposition method, although sputtering may be used in some situations. The caveat is that the deposited thin film is disconnected at the PR openings owing to their re-entrant profiles. After that, the substrate is immersed in a solution that dissolves the PRs and, during the process, removes the excessive deposited thin film material on the PR (Figure 3.10c). Sometimes ultrasonic agitation is needed to ensure PR removal.

3.6.2 Annealing

Annealing technique is used mostly in IC microfabrication after the introduction of dopants to the substrate. The purpose is to repair the damage to the substrate and activate the dopants. Annealing can also densify the deposited film and release or modify the mechanical stress within the thin films intended for mechanical structures. These changes may be critical for the operation of MEMS devices.

Annealing is generally performed in a furnace at high temperatures in the presence of nitrogen. The process takes tens of minutes to hours. Rapid thermal annealing (RTA) is an alternative that takes only seconds to minimize diffusion in the otherwise long furnace annealing.

3.6.3 Liquid Phase Photopolymerization

The liquid phase photopolymerization process is used to photopattern polymer structures. The essence of this process is the same as the photolithography process described earlier. The main difference is that the liquid form prepolymer is not spin coated onto the substrate but rather is filled into a predefined cavity or channel. Like negative PRs, the prepolymers are photosensitive, and upon masked exposure to light (mostly UV), they undergo polymerization and solidify. Unexposed pre-polymer will not polymerize and can be rinsed away.

This process can apply to many polymerizable materials such as PDMS and IBA, and is especially useful for rapid prototyping of microfluidics and lab-on-a-chip devices. Figure 3.11 shows an example of patterning an IBA structure using this technique. Many options can be used to form the cavity to fill the prepolymers. For example, a cartridge with spacers along its edges is utilized in Figure 3.11a. By fixing the cartridge onto the substrate, a cavity is formed. Then the photosensitive IBA prepolymer can be filled into the cavity. Next, a masked UV exposure is carried out to pattern the IBA (Figure 3.11b). After the UV-exposed IBA solidifies, the cartridge can be peeled off (Figure 3.11c). Finally, any unpolymerized IBA prepolymer is rinsed away with ethanol, and the desired IBA polymer structure is made on the substrate (Figure 3.11d).

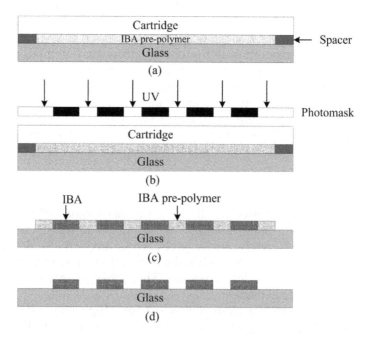

FIGURE 3.11
Liquid phase photopolymerization.

3.6.4 Micromolding

Micromolding is a template replication process in which the substrate serves as a mold and is not part of the finished device. The greatest benefit of this process is that the mold can be reused many times, reducing overall manufacturing cost. Conventional methods include *injection molding* (molten plastics are injected into metal molds, cooled, and removed) and *hot embossing* (a thermoplastic material is inserted into a modeling machine and formed under pressure). For the formation of microscale molds, microfabrication techniques discussed above can be utilized. For example, Si wafers can be etched to render the desired molds. Another example is using liquid phase photopolymerization of PDMS or IBA (described above) to form micromolds.

3.6.5 Soft Lithography

Soft lithography is an alternative way of transferring patterns onto a substrate [13]. Unlike photolithography, this technique utilizes a polymeric mold, often made of PDMS, for direct physical transfer. The PDMS may be first coated with the material to be patterned, followed by stamping onto the substrate. The desired material is thus patterned onto the substrate following the raised parts of the PDMS stamp. This process, called *microcontact printing*, is similar to pressing a stamp onto an ink pad and then onto a piece of paper. The stamped material may also be utilized as a masking component for etching into a substrate.

The relief parts of the PDMS stamp can also be used to transfer patterns. In this approach, the PDMS stamp is first filled with a polymer precursor and then pressed against the substrate. After curing (solidifying) the polymer, the PDMS stamp is removed and a pattern matching its relief parts remains on the substrate. This process is called *microtransfer molding*. Many other processes implement similar concepts.

3.6.6 Electroplating

Electroplating is a useful process to produce thick metal structures on a substrate. Many metals such as copper (Cu), Au, and Ni can be electroplated. The substrate to be electroplated is immersed in an electroplating solution that contains a reducible form of the ion of the desired metal. The substrate is maintained at a negative potential (cathode) relative to an inert positive counter electrode (anode; e.g., platinum). During the electroplating, electrons are supplied to the surface of the exposed metallic regions on the substrate, and the metal ions are reduced to their atomic form and subsequently deposited onto the substrate surface.

The surface of the substrate to be electroplated must first be metalized to allow currents to flow. This is usually done by sputtering a thin, submicron layer of metal (seed layer) onto the surface. Multiple thin layers of metal may

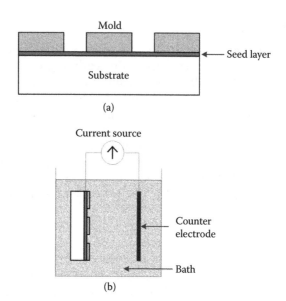

FIGURE 3.12
Electroplating process. (a) Molds are first formed on the substrate to define regions to be electroplated. (b) Electroplating setup.

be sputtered to enhance the adhesion of the plated metal to the surface of the substrate. If only selected regions on the substrate surface are to be plated, a mold is first patterned onto the substrate through photolithography or other approaches described earlier. The open regions will be plated and the areas occupied by the mold will not. The mold and the seed metal layers covered by the mold may or may not be removed and etched away after the electroplating, depending on the application. Electroplating can also be applied to certain photoresists [14]. Figure 3.12 shows electroplating.

3.6.7 Sacrificial Process

Sacrificial process is an extremely important concept for creating movable structures on a substrate. Structural materials must be anchored onto the substrate but a gap must be left between the structural material and the substrate to allow motion of the structure. Since we cannot deposit a structural material suspended in air, the ultimate gap must be occupied initially by a material that will be removed eventually. It is known as a *sacrificial* material and its removal is called *release*.

Figure 3.13 illustrates how a polycrystalline silicon cantilever beam is formed onto a Si substrate. We first deposit a layer of SiN onto the Si substrate as an insulator. Next, a layer of SiO_2 is deposited and patterned onto the substrate via LPCVD, followed by photolithography and masked etching of SiO_2 to expose the underlying SiN at a designated anchor area. Polycrystalline

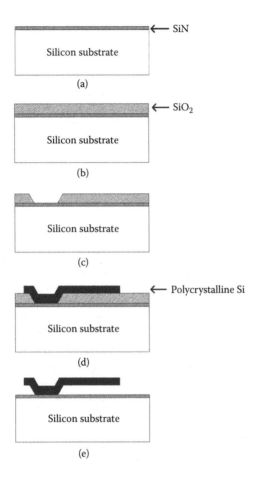

FIGURE 3.13
Sacrificial process to create a polycrystalline silicon cantilever beam. (a) SiN layer is deposited onto Si substrate as insulating layer. (b) SiO$_2$ layer is deposited as sacrificial layer. (c) SiO$_2$ is photopatterned and etched to expose anchoring area for future cantilever beam. (d) Polycrystalline Si layer is deposited, photopatterned, and etched to form cantilever beam. (e) Cantilever beam is released after removal of SiO$_2$ sacrificial layer.

silicon is subsequently deposited via LPCVD, and patterned and etched to form the cantilever.

Note that the SiO$_2$ layer underneath serves as a place holder for the ultimate air gap between the polycrystalline silicon cantilever and the substrate. Now we can release the cantilever by etching away the sacrificial SiO$_2$. This is done in a wet etching step utilizing hydrofluoric acid (HF).

3.6.8 Bonding

The bonding process mechanically attaches similar to dissimilar substrates together, either permanently or temporarily. In some cases, electrical connection

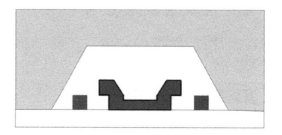

FIGURE 3.14
Hermetically sealed structures. Devices are fabricated on one substrate that is bonded to another with a cavity.

between the two substrates is also fulfilled. Bonding is especially useful when devices are enclosed in a cavity and must be sealed from the outside environment to prevent contamination, damage, or other effects on operation such as moisture and oxidation. Hermetic sealing is one example of sealing (Figure 3.14).

One basic type is *anodic bonding* that can bond metal to glass, Si to glass, or Si to Si with an intermediate glass layer. The glass layer can be sputtered, spun on, or even evaporated. Anodic bonding requires a high temperature or strong electric field to form metal oxides and mix them with glass. The second method is *fusion bonding* that involves direct Si to Si bonding. The two Si surfaces may or may not contain thermally grown SiO_2. The third approach is simply using adhesives to "glue" the two substrates together. Examples of adhesives are PRs, silicone rubber, and polyepoxide (epoxy) resins.

3.6.9 Surface Modification

Surface modification is widely used to change the properties of a surface to suit specific applications. It has found increasing use in biotechnology applications. Our discussion on this topic is limited to the modification of surface hydrophilicity and hydrophobicity. As discussed in Chapter 2, they are very important factors in the creation of liquid microlenses.

The first way of modifying the surface hydrophilicity and hydrophobicity is to apply an ultrathin monolayer of chemical onto the surface. This is often referred to as a self-assembled monolayer (SAM). One example in photolithography is to coat hexamethyldisilazane (HMDS) onto the substrate surface to enhance the adhesion of PR. Another example is applying octadecyltricholorosilane (OTS) onto a glass substrate to make it hydrophobic.

The second method involves plasma treatment of the surface. For example, when two plasma oxidized PDMS surfaces are brought into conformal contact, an irreversible seal forms between them. This technique can be used to bond two PDMS layers [15,16]. Oxygen plasma treatment of PDMS also enables the bonding of PDMS to other materials such as glass [17]. Plasma oxidation of a PDMS surface makes it more hydrophilic from its native hydrophobic state [15,16].

The third approach is modifying the surface morphology. For example, super-hydrophobicity (contact angle approaching 180 degrees) may be accomplished when a surface is covered with nanoscale fabricated structures [18].

3.6.10 Laser-Assisted Processes

Laser beams with sufficient power can be used for different processes. *Laser drilling* is a technology for inserting holes into or through Si wafers, glass, or other substrates through ablation. The method does not have specific requirements for ambient gases. *Laser annealing* is similar but uses lower energy and usually produces larger spot sizes. The heat generated by a laser is used to anneal the area covered by the spot.

Laser beams can also be used to drive chemical reactions in a technique known as *laser-assisted chemical etching*. Much higher local etch rates can be achieved with this method than are possible with RIE. Similarly, a laser beam can supply energy to drive a deposition; the method is called *laser-driven deposition*. An example is depositing W onto specific areas of a substrate from tungsten hexacarbonyl [$W(CO)_6$]. Laser-assisted processes can be used to create interesting three-dimensional structures. However, the processes are intrinsically serial, not parallel, meaning that a beam must be scanned across the areas to be processed spot by spot instead of simultaneously.

3.6.11 Planarization

After multiple fabrication steps, an originally flat wafer surface develops a topography. Such topography is, of course, essential to the creation of the desired device structures. Nonetheless, it also presents difficulties during subsequent processing steps such as uniform thin film deposition, photolithography, and etching.

To address this issue, planarization of the wafer surface can be applied. One approach is *mechanical polishing*, in which the wafer is pressed against a rotating platen and polished by an abrasive slurry. If the slurry includes chemicals that help dissolve the removed material, the process is called *chemical mechanical polishing*.

The second approach is spin casting polymers onto a substrate surface. Examples include using polyimides and a PR called SU-8. The third method is *resist etchback*. The wafer is first spin coated with a PR to planarize the surface, followed by a blank plasma etch that has the same approximate etching rate of the PR as that of the underlying material (e.g., SiO_2) and produces a much more planar surface.

3.6.12 Fabrication on Flexible Substrates and Curved Surfaces

Conventional microfabrication technologies apply only to flat and rigid substrates and materials. In recent years, advances in technologies allow fabrication on flexible substrates and curved surfaces. Flexible technology is an active research area. We will briefly discuss a few examples.

The first one essentially reinvents the microfabrication processes by replacing the rigid solid materials with flexible polymer counterparts in semiconductors, conductors, and insulators. The second approach is to thin the films used to make the devices, followed by transfer onto flexible substrates. For example, nanometer-thick Si is flexible enough to allow transfer onto polymer substrates such as PDMS while maintaining its superb properties as a semiconductor [19].

The third approach takes full advantage of established microfabrication techniques [20,21]. The devices and systems are still fabricated onto traditional substrates but divided into smaller parts that reside on individual small islands. These islands are connected mechanically, electrically, or both ways by thin flexible polymer substrates. The complete structure may then be transferred onto other flexible structures or curved surfaces. Most of the mechanical stress suffered from such transfers will be absorbed by the thin polymer substrates; the small islands and the devices on them experience much less stress.

3.7 Examples of Microfabrication Processes

We present three detailed examples of microfabrication to help the readers better understand how a fabrication flow for making devices is constructed. We have endeavored to incorporate all the basic fabrication steps as well as many techniques described in this chapter. Note that these examples were chosen intentionally to better prepare readers for subsequent chapters about microlenses rather than explain conventional IC fabrication. For more insight into IC fabrication, readers are referred to many excellent books that focus on the subject.

3.7.1 Example 1

Our goal is to fabricate a 100-μm-tall microscale magnetic stirrer made of Ni in a microfluidic channel filled with water. The stirrer can spin under the influence of a rotational magnetic field. Figure 3.15 shows the fabrication process flow.

Step 1: The choice of substrate is not specified, but Si or glass will work. We intend to use electroplating to form the Ni stirrer. Hence, the first step is to sputter the seed layers for the electroplating. We sputter three layers of metal thin films in sequence (Ti, Cu, and Ti), each about 20 nm thick. The bottom Ti layer is to enhance the adhesion of the future electroplated Ni onto the substrate. The middle layer of Cu is the seed layer for Ni electroplating.

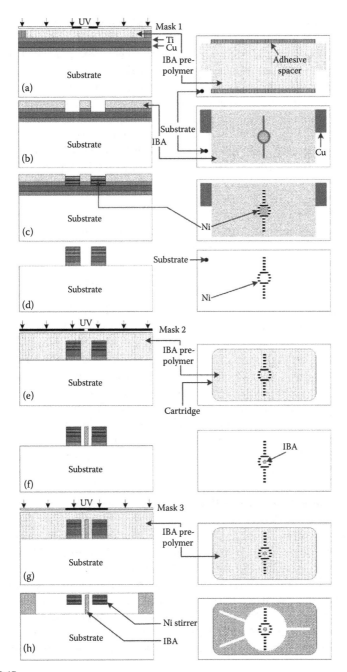

FIGURE 3.15
Fabrication process flow of Ni stirrer in microfluidic channel. The left column shows the cross section and the right column the top view. (*Source:* Adapted from Agarwal, A.K., Sridharamurthy, S.S., Beebe, D.J. et al. 2005. *Journal of Microelectromechanical Systems*, 14(6), 1409–1412. With permission.)

Since Cu is easy to oxidize in air, the top Ti layer serves to protect Cu from oxidation. See Figure 3.15a.

Step 2: We will need a mold to perform the electroplating. The thickness of this mold must exceed the desired height of the electroplated Ni stirrer (100 μm). Hence we choose to utilize liquid phase photopolymerization rather than a PR since it is easier to generate structures with heights on this order.

We first tape the substrate at the edges with an adhesive spacer. The height of the adhesive spacer will also define the height of the mold, say 150 μm, satisfying the prerequisite of exceeding the height of the Ni stirrer. Then we cover the substrate with a piece of transparency on which patterns have been printed to serve as the photomask. The photomask along with the substrate and the adhesive spacers form a chamber.

We fill the chamber with a photosensitive prepolymer, for example, IBA. After a UV exposure with the photomask, most of the IBA is polymerized and the rest is rinsed away, leaving behind on the substrate an opening bearing the shape of the stirrer. Other regions of the substrate are covered with IBA. Finally, the photomask is removed to expose the IBA mold. See Figures 3.15a and b.

Step 3: Electroplating can now be carried out. Note that the metal layers under the PR mold are all connected, covering the whole substrate. Hence the electrode can be connected to the edge of the metal. However, keep in mind that the top Ti protection layer must be removed first to expose the Cu seed layer for the plating of Ni. This can be done in a wet etching step. Next Ni is plated to a thickness of 100 μm, the desired height of the stirrer. The areas covered by the IBA mold are not plated; see Figure 3.15c.

Step 4: After the electroplating, we remove the IBA mold with ethanol and use one more wet etching step to etch away the Ti, Cu, and Ti layers. However, the plated Ni is still anchored onto the substrate through the Ti and Cu layers beneath that are mostly intact in this wet etching step. This is intentional because we cannot release the stirrer before completing other fabrication steps. See Figure 3.15d.

Step 5: We now define the axis around which the stirrer will spin and the channels where water will be held and where the stirrer resides. This involves two steps of liquid phase photopolymerization of IBA in sequence using two photomasks. The first one defines the axle around which the stirrer will revolve. The second one forms the microfluidic channel housing the stirrer. This time we use cartridges to form the chambers on the substrate in which the IBA prepolymer is filled. After the first photopatterning in this step, the cartridge is removed. However, after the second photopatterning, the cartridge can remain as the top of the microchannel. See Figure 3.15e through h.

Step 6: We release the stirrer. We fill the channel with etchants to wet etch the remaining Ti and Cu beneath the Ni stirrer, so that it becomes floating. Note that the Ti and Cu beneath the Ni stirrer served as sacrificial materials.

Finally, we rinse the etchants away and refill the channel with water. See Figure 3.15h. If we apply a rotational magnetic field beneath the substrate, the Ni stirrer will spin!

In this example, we observed that a fabrication process flow involves multiple steps that depend closely on each other, from the choice among candidate methods to the determination of the process parameters. In fact, the design of the complete process flow is an iterated procedure and inevitably requires trial and error.

3.7.2 Example 2

We intend to fabricate high aspect ratio Si structures onto a flexible PDMS film. The overall strategy is to fabricate the structures through DRIE first, followed by their transfer to PDMS (Figure 3.16). The process is illustrated in Figure 3.17.

Step 1: We start by choosing a suitable substrate—a silicon-on-insulator (SOI) wafer. An SOI wafer consists of a thin (1 μm or less) SiO_2 layer sandwiched between two Si layers. We first thermally grow 1 μm of SiO_2.

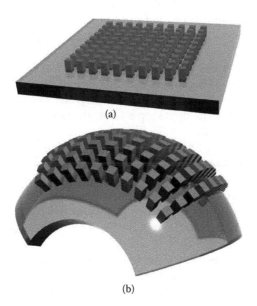

(a)

(b)

FIGURE 3.16

Fabrication of high aspect ratio Si structures onto flexible polymer substrate. The intent is to fabricate planar Si structures first and transfer them onto a polymer substrate (a) so that they can lie on a curvilinear surface (b). (*Source:* Zeng, X. and Jiang H. 2011. *Journal of Microelectromechanical Systems*, 20(1), 6–8. With permission.)

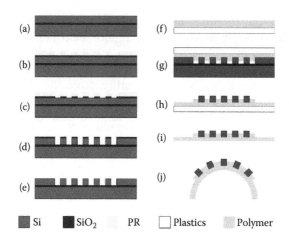

FIGURE 3.17
Fabrication process flow for structures shown in Figure 3.16. (*Source:* Zeng, X. and Jiang, H. 2011. *Journal of Microelectromechanical Systems*, 20(1), 6–8. With permission.)

This layer will serve as the masking material for the deep Si etching. See Figure 3.17a.

Step 2: We then transfer the designed Si structures onto the SiO_2 layer. A photolithography step is carried out for this purpose, and the SiO_2 layer is wet etched with buffered HF. The etch stops at the underlying Si. See Figures 3.17b and c.

Step 3: The PR is removed. Next DRIE is performed to etch the Si now exposed using the SiO_2 on top as the masking material until the buried SiO_2 is reached. The first reason for using an SOI wafer becomes clear. The buried SiO_2 layer serves as the etch stop so that the Si below will not be further attacked. See Figure 3.17d.

Step 4: The remaining SiO_2 masking material is removed by buffered HF wet etching (Figure 3.17e).

Step 5: We transfer the whole structure to a polymer substrate. First we spin coat a thick layer of PR (SU-8) onto a plastic sheet. Next we attach the plastic sheet with SU-8 onto the etched SOI wafer. Then a blank (non-masked) UV exposure (flood exposure) cures the SU-8 to "grab" the Si structures. See Figures 3.17f and g.

Step 6: We release the high aspect ratio Si structures. We place the plastic sheet with the attached Si structures in buffered HF again to remove the buried SiO_2 under the Si structures. The second reason to choose an SOI wafer has been revealed. The buried SiO_2 also serves as the sacrificial material. After this step, the Si structures formed by DRIE are released and transferred onto the substrate of SU-8 and plastic. See Figure 3.17h.

Step 7: Finally, we can remove the plastic handler. The flexible SU-8 film with the Si structures transferred onto it can now deform to a curvilinear shape. See Figures 3.17i and j.

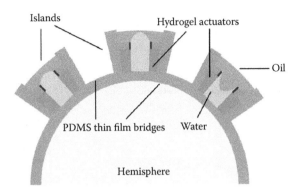

FIGURE 3.18
Multiple islands arranged on hemisphere. Each island houses a tunable liquid microlens formed by a curved water-to-oil interface pinned at the edge of a polymer aperture by surface tension. The change in the curvature of the interface is caused by stimulus-responsive hydrogel actuators. (*Source:* Adapted from Zhu, D., Zeng, X., Li, C. and Jiang, H. 2011. *Journal of Microelectromechanical Systems*, 20(2), 389–395. With permission.)

3.7.3 Example 3

Our final example involves more extensive fabrication steps. The goal is to fabricate a polymer structure that can be wrapped onto a hemisphere. On the polymer structure are individual islands where tunable liquid microlenses based on pinned, curved water-to-oil interfaces ultimately will reside. The general strategy is to connect these islands by bridges made of a thin polymer film so that when the whole structure is wrapped onto a hemisphere, most of the mechanical stress will be absorbed by the thin polymer film instead of affecting the islands. We will pin the liquid lenses utilizing surface tension; therefore, surface modifications to render different regions more hydrophobic or more hydrophilic are also necessary. Multiple molding steps are required to form the islands. Figure 3.18 shows the intended device structure. Figure 3.19 outlines the main steps of the fabrication process.

Step 1: The first layer of the first IBA mold for the PDMS islands is patterned using liquid phase photopolymerization. See Figure 3.19a. The IBA layer is photopatterned on a glass slide with Mask I, which is shown on the right of Figure 3.19a.

Step 2: The second layer of the first IBA mold is formed on top of the first layer of the IBA after Step 1. See Figure 3.19b. This is another liquid phase photopolymerization process using Mask II, shown on the right. Mask II is aligned with the first IBA layer by three small circles on the edge. By heightening specific parts of the IBA layer in this second lithography step, molds for the connection bridges are realized.

Step 3: PDMS serving as the substrate and the connection bridges is formed using the first IBA mold. See Figure 3.19c. The top view of the geometry of the PDMS pattern is shown on the right.

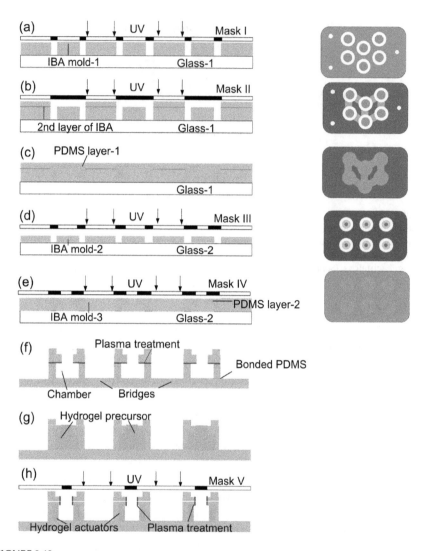

FIGURE 3.19

Fabrication process flow for device described in Figure 3.18. (*Source:* Adapted from Zhu, D., Zeng, X., Li, C. and Jiang, H. 2011. *Journal of Microelectromechanical Systems,* 20(2), 389–395. With permission.)

Step 4: The second IBA mold is photopatterned on a second glass slide with Mask III. See Figure 3.19d. The top view of the geometry is shown on the right.

Step 5: The third IBA mold is photopatterned with Mask IV. This IBA mold will later define the PDMS barriers for the liquids. Then the PDMS structure is formed using the IBA molds. This PDMS structure will serve as the aperture layer of the microlenses. See Figure 3.19e. The top view of the geometry is shown on the right.

Step 6: The two PDMS layers are stripped from their molds and bonded together. Plasma treatment is performed on their surfaces to facilitate the bonding. The bonded PDMS structures form the chambers to house the liquid microlenses. See Figure 3.19f.

Step 7: Hydrogel precursor is injected into the chambers, followed by another liquid phase photopolymerization step to define the actuators. The sidewalls of the lens apertures are plasma treated to be more hydrophilic. See Figures 3.19g and h. The whole structure is now ready to be wrapped onto the target hemisphere for final assembly of the microlenses.

References

1. S. M. Sze, *VLSI Technology*. New York: McGraw Hill, 1988.
2. C. Y. Chang and S. M. Sze, *ULSI Technology*. New York: McGraw Hill, 1996.
3. S. A. Campbell, *The Science and Engineering of Microelectronic Fabrication*, 2nd ed. New York: Oxford University Press, 2001.
4. P. van Zant, *Microchip Fabrication: A Practical Guide to Semiconductor Processing*, 4th ed. New York: McGraw Hill, 2000.
5. J. D. Plummer, M. D. Deal, and P. B. Griffin, *Silicon VLSI Technology: Fundamentals, Practice and Modeling*. Upper Saddle River, NJ: Prentice Hall, 2000.
6. J. W. Gardner, V. K. Varadan, and A. O. Awadelkarim, *Microsensors, MEMS, and Smart Devices*. Chichester: John Wiley & Sons, 2001.
7. S. Franssila, *Introduction to Micro Fabrication*. Chichester: John Wiley & Sons, 2004.
8. S. D. Senturia, *Microsystem Design*. Norwell, MA: Kluwer Academic, 2001.
9. G. T. A. Kovacs, *Micromachined Transducers Sourcebook*. New York: McGraw Hill, 1998.
10. C. Liu, *Fundamentals of MEMS*, 2nd ed. Upper Saddle River, NJ: Prentice Hall, 2012.
11. S. S. Saliterman, *Fundamentals of BioMEMS and Medical Microdevices*. Bellingham, WA: SPIE, 2006.
12. L. J. Chen, "Metal silicides: an integral part of microelectronics," *JOM*, vol. 57, pp. 24–31, 2005.
13. Y. N. Xia, and G. M. Whitesides, "Soft lithography," *Annual Review of Materials Science*, vol. 28, pp. 153–184, 1998.
14. P. Kersten, S. Bouwstra, and J. W. Petersen, "Photolithography on micromachined 3D surfaces using electrodeposited photoresist," *Sensors and Actuators A*, vol. 51, pp. 51–54, 1995.
15. D. C. Duffy, J. C. McDonald, O. J. A. Schueller, and G. M., Whitesides, "Rapid prototyping of microfluidic systems in poly(dimethylsiloxane)," *Analytical Chemistry*, vol. 70, pp. 4974–4984, 1998.
16. M. A. Eddings, M. A. Johnson, and B. K. Gale, "Determining the optimal PDMS–PDMS bonding technique for microfluidic devices," *Journal of Micromechanical Microengineering*, vol. 18, p. 067001, 2008.

17. S. Bhattacharya, A. Datta, J. M. Berg, and S. Gangopadhyay, "Studies on surface wettability of poly(dimethyl) siloxane (PDMS) and glass under oxygen plasma. treatment and correlation with bond strength," *Journal of Microelectromechanical Systems*, vol. 14, pp. 590–597, 2005.

18. T. Onda, S. Shibuichi, N. Satoh, and K. Tsujii, "Super water-repellent fractal surfaces," *Langmuir*, vol. 12, pp. 2125–2127, 1996.

19. M. G. Lagally, "Silicon nanomembranes," *MRS Bulletin*, vol. 32, pp. 57–63, 2007.

20. H. C. Ko, M. P. Stoykovich, J. Song, V. Malyarchuk, W. M. Choi, C.-J. Yu, J. B. Geddes III, J. Xiao, S. Wang, Y. Huang, and J.A. Rogers, "A hemispherical electronic eye camera based on compressible silicon optoelectronics," *Nature*, vol. 454, pp. 748–753, 2008.

21. X. Zeng and H. Jiang, "Fabrication of complex structures on non-planar surfaces through a transfer method," *Journal of Microelectromechanical Systems*, vol. 20, pp. 6–8, 2011.

4

Solid Microlenses

This chapter focuses on examples of microlenses made of solids whose focal lengths cannot be tuned during their operations. Microlenses with fixed focal lengths whose focal points are shifted by moving their positions are also discussed in this chapter. These non-tunable microlenses include:

Ge/SiO$_2$ core/shell nanolenses

Microlens arrays fabricated through molding processes

Injection-molded plastic lenses

Lenses formed by thermally reflowing photoresists (PRs)

Self-assembled supermolecular nanoscale spherical lenses

Lenses made by graded exposure in PR

Lenses fabricated by direct photo-induced polymerization

Strain-responsive lens arrays

Lenses latched by electrowetting

Lens arrays fabricated from melting polystyrene beads

Lens arrays fabricated by dewetting methods

Lenses formed from inkjet printing

4.1 Introduction

Fixed focal length microlenses and arrays with lens diameters of a few to several hundred micrometers are extensively used in many optical systems such as projection lithography [1–3], optical coherence microscopy (OCM) [4], Shack-Hartmann sensors [5], and back lighting for projection liquid crystal displays (LCDs) [6].

In this chapter, we will cover the various lens fabrication techniques. The lens materials range from polymers to glass and include polycarbonates, ultraviolet (UV)-curable epoxies, polyimides, and polystyrenes. These techniques can produce lenses with numerical apertures (*NAs*) less than 0.6, typically 0.1 to 0.3 [7]. Low *NA* values are attributable to the surface profiles and refractive indices of microlenses.

4.2 Germanium and Silicon Oxide Nanolenses

The first lens described here is a nanolens made of germanium (Ge) and silicon dioxide (SiO_2) that may potentially satisfy the growing need to integrate electronic components with optoelectronics for telecommunications and connection of computers [8]. Because they exhibit narrower bandgaps than silicon, silicon germanium heterostructures have been used for various optical devices such as modulators and photodectectors. Taking advantage of the comparably low etching rate and uniform size of self-assembled Ge quantum-dot (QD) multiple layers, Chen et al. developed a method to fabricate multilayered Ge/SiO_2 core/shell structures for nanolenses with excellent uniformity over a large area [8].

Figure 4.1 depicts the formation of multilayered Ge/SiO_2 nanolenses [8]. Deposition was first performed in an ultrahigh vacuum chemical vapor deposition (UHVCVD) apparatus. Then the sample with Ge QDs on top, as seen in Figure 4.1a, was wet etched by aqueous tetramethylammonium hydroxide (TMAH) solutions. The etching rate of the Si–spacer layer was much faster than that of the Ge QDs. Therefore, the uppermost monodispersed Ge QDs could keep their shapes. As the etching further proceeded on the uncapped Si surface, V-shaped grooves caused by anisotropic wet etching appeared, as shown in Figure 4.1b.

The Ge QDs were gradually stacked while coated with amorphous SiO_2 shells. Eventually, multilayered Ge/SiO_2 core/shell nanolenses were thus formed, as seen in Figure 4.1c.

Figure 4.2a shows a cross-sectional transmission electron microscopy (XTEM) image of the fabricated multilayered Ge/SiO_2 nanolenses [8]. The Ge QDs had an average diameter and density of 70 ± 5 nm and 4.5×10^9 cm^{-2}, respectively. A top view field-emission scanning electron microscopy image is shown in Figure 4.2b.

Figure 4.3(a) shows the photoluminescence (PL) spectrum at 10 K of multilayered Ge nanolenses prepared with 45 seconds (s) of immersion in TMAH solution. The broad peak with energy at 0.79 eV was due to the multilayered nanolenses and could be deconvolved into two Gaussian-like peaks at 1.5 and 1.55 µm, respectively. Multilayered nanolenses consist of aggregations of SiO_2 nanoparticles at the surfaces of Ge QDs with core/shell structures. The authors assumed that for their multilayered nanolenses, the indirect bandgap *Eg* resembles that of Ge and is given by the energy difference between the L conduction band minimum and the top of the valence band. Owing to the atomic intermixing during the growth of multilayered Ge QDs, the bandgap energy increases with the intermixing Si composition.

The reflection spectra of the multilayered Ge nanolenses were measured using a Bomem Fourier transform infrared (FTIR) spectrometer (Figure 4.3b). The reflection spectra exhibited two broad peaks centered at wavelengths of 1.2 and 1.5 µm, respectively. The significant increase in reflectivity at 1.5 µm for samples with Ge–nanolens stacks indicated that they could be potentially useful for photodetectors operating in the telecommunication range [8].

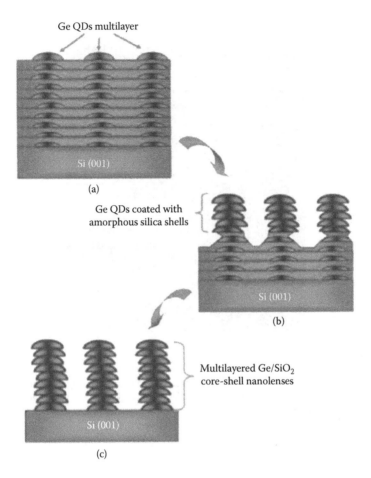

Ge QDs multilayer

(a)

Ge QDs coated with amorphous silica shells

(b)

Multilayered Ge/SiO₂ core-shell nanolenses

(c)

FIGURE 4.1
Formation of multilayered Ge/SiO$_2$ nanolenses. (a) As-grown layered structure. (b) Initial stage of etching that introduces V-shaped grooves. (c) Final structure with multilayered Ge/SiO$_2$ nanolenses. (*Source:* Chen, H.C., S.W. Lee, and L.J. Chen. 2007. *Advanced Materials*, 19(2), 222–226. With permission.)

4.3 Quartz Glass Microlenses Etched by Reactive Ion Etching

Unlike the previous example, a microlens etched by a dry technique on quartz glass is shown in this section. In optical communications, single mode fibers are preferred because of the absence of mode dispersion and mode noise. In order to explore a microlens whose working wavelength lies in the infrared (IR) frequency band (e.g., 1300 nm), Eisner et al. presented silicon microlenses for IR fabricated by reactive ion etching (RIE) [9].

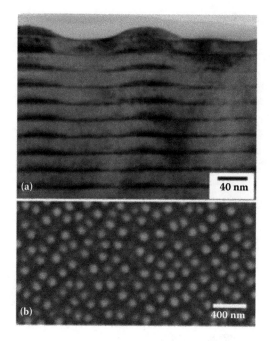

FIGURE 4.2
(a) Cross-sectional transmission electron microscopy (XTEM) image showing as-prepared self-assembled Ge-QD/Si-spacer multilayer structure. (b) Top view scanning electron microscopy (SEM) image of Ge QDs. (*Source:* Chen, H.C., S.W. Lee, and L.J. Chen. 2007. *Advanced Materials,* 19(2), 222–226. With permission.)

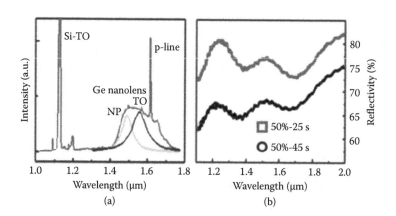

FIGURE 4.3
(a) Photoluminescence spectra at 10 K of a sample etched in TMAH solution for 45 seconds. (b) Reflection spectra of samples etched in TMAH solution for 25 and 45 seconds with signals from nonetched samples subtracted. (*Source:* Chen, H.C., S.W. Lee, and L.J. Chen. 2007. *Advanced Materials,* 19(2), 222–226. With permission.)

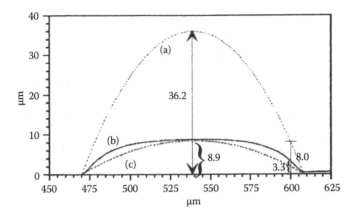

FIGURE 4.4

Measurement of profile of quartz glass lens with mechanical profilometer. Curve (a) shows a PR lens etched into the quartz glass, curve (b). Curve (c) is an ideal sphere at identical height to the glass lens. (*Source:* Eisner, M. and J. Schwider. 1996. *Optical Engineering*, 35(10), 2979–2982. With permission.)

The fabrication of this lens takes advantage of the nature of RIE. As noted in Chapter 3, RIE involves both ion bombardment and chemical reaction. Because the incident direction of ions was vertical to the cathode, the number of ions per unit area (ion density) was greater at the center of an opening than in its rim region. This fact contributed to a stronger etching in the middle than at the edge. In other words, the etch rate decreased from the center to the edge as a result. Figure 4.4 shows the profile of an etched quartz glass microlens measured by a mechanical profilometer. The dotted curve (a) represents the spherical PR lens (Section 4.9 discusses PR microlenses). Curve (b) is the profile of the etched glass lens and shows clearly that the rim is steep and the center is quite flat, compared with the ideal spherical shape shown by curve (c) at the identical height.

A lens produced in this manner shows strong spherical aberrations. To produce a better lens, a proper choice of etching parameters is critical and would generate profiles close to elliptic or hyperbolic forms. For this purpose, the authors further explored the etching selectivity that could be changed by varying the etching parameters (gas flow, gas mixture, and power) during the etching process [9]. The resultant lens after etching into silicon was much improved; the interferogram is shown in Figure 4.5. A deviation in profile of only 0.24 λ (λ = 633 nm) was achieved from a reference sphere. Hence, an "ideal" single lens shape could be approximated.

Roughness is a serious problem for all dry etching techniques. Eisner and Schwider [9] achieved better surface roughness by adding the noble helium gas to the etching gas. Helium has a high thermal conductivity and provides some additional cooling for the wafer. Figure 4.6 shows an SEM image and atomic force microscopy (AFM)-scanned surface of a microlens, indicating a very smooth surface.

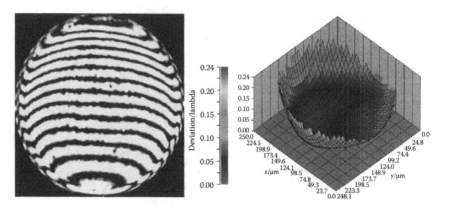

FIGURE 4.5
Twyman-Green interferogram and three-dimensional plot of silicon microlens. (*Source:* Eisner, M. and J. Schwider. 1996. *Optical Engineering,* 35(10), 2979–2982. With permission.)

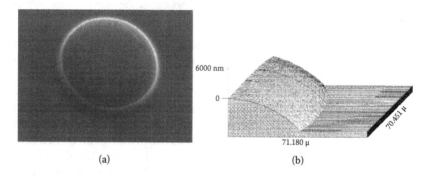

(a)   (b)

FIGURE 4.6
(a) Scanning electron microscopy image of microlens. (b) Atomic force microscopy image of microlens edge. The surface of the microlens is very smooth. (*Source:* Eisner, M. and J. Schwider. 1996. *Optical Engineering,* 35(10), 2979–2982. With permission.)

4.4 Self-Assembled Supermolecular Nanoscale Spherical Microlenses

Lee et al. show a self-assembly method to form nanolenses made of calix[4]hydroquinone (CHQ). This compound is composed of four p-hydroquinone subunits with eight hydroxyl groups [10]. Dissolving the CHQ monomers in 1:1 water-acetone solution leads to the formation of needle-like CHQ nanotube crystals with infinitely long hydrogen-bonded arrays. When the crystals are heated at 40°C in aqueous environments, CHQ molecules released from the crystals can re-assemble into nanospheres.

These planospherical convex (PSC) structures (spherical face on one side and a flat face on the other) can be isolated as nanolenses. CHQ lenses are

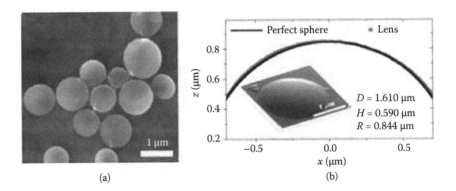

(a) (b)

FIGURE 4.7
CHQ planospherical convex lenses. (a) SEM image showing various sizes of CHQ lenses separated as an aqueous suspension and drop-dried on a substrate. (b) AFM profile showing the near perfect spherical face of the lens. (*Source:* Lee, J.Y. et al. 2009. *Nature*, 460(7254), 498–501. With permission.)

stable in air and may be utilized for bioimaging, near field lithography, optical memory storage, light harvesting, spectral signal enhancing, and optical nanosensing [10]. The size distribution of CHQ lenses can be controlled by the time and temperature of the self-assembly process.

Figure 4.7a shows CHQ nanolenses of various sizes. Typically, PSC lenses with nanoscale thickness [height (H) < 800 nm and diameter (D) between 0.05 and 3 μm] can be synthesized and separated from the aqueous suspension. The round surfaces of the nanolenses exhibited a deviation from a spherical surface of less than 3% with a surface roughness less than 1 nm, as shown in Figure 4.7b.

Figure 4.8 shows an optical microscope image of a CHQ lens (D = 970 nm, H = 220 nm) on top of CHQ tubule bundles under filtered light (λ_{max} = 472 nm) from a halogen lamp. The image indicates that the lens magnified the underlying object with a magnification factor of 1.6. The paraxial focal length estimated from the observed magnification was calculated at 590 nm—much shorter than that predicted using geometric optics (1.3 μm). The deviations from geometrical optics for these CHQ lenses stem from their sub-wavelength size and are the key features of near field focusing [10].

To demonstrate enhanced spatial resolution, the authors [10] also investigated the optical properties of CHQ lenses on pre-fabricated sub-wavelength objects, as shown in Figure 4.9. The Rayleigh resolution limit for point objects was 320 nm, while that for line objects was 262 nm. A more stringent Sparrow resolution limit was 249 nm.

The optical images outside the CHQ lenses did not resolve the underlying stripe patterns, as the stripe spacings of 220 and 250 nm were narrower than or very similar to the stringent resolution limit. Nonetheless, the stripes could be clearly imaged and resolved through a CHQ lens (Figure 4.9a and c for 250 nm and e for 220 nm).

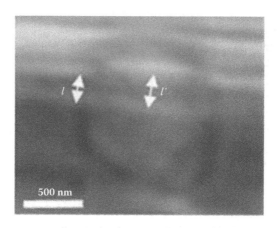

FIGURE 4.8
Optical microscope image of a CHQ lens on a CHQ nanotube crystal, showing the magnification by the lens. The line spacing (*l*) behind the lens is considerably increased (*l'*). (*Source:* Lee, J.Y. et al. 2009. *Nature,* 460(7254), 498–501. With permission.)

The image magnification increased as the distance between nanolens and image increased. The magnifying effect enhanced the resolution substantially—as much as 2.5 times. The magnified images of the face-up lenses showed pincushion distortion; no notable distortion appeared for the face-down lens. This difference could be attributed to the fact that the near field image of the face-up lens was formed by the interference of secondary Fresnel waves on the flat and convex surfaces of the lens while the near field image of the face-down lens was formed by the secondary surface waves due to the convex surface.

4.5 Microlens Arrays Fabricated from Self-Assembled Organic Polymers

Hayashi et al. applied a self assembly of polystyrene beads to form two-dimensional (2D) lattices on glass substrates, and demonstrated their use as arrays of microlenses in imaging [11]. However, this method could provide only one array with one specific pattern. In addition, high quality images are hard to obtain from microlenses shaped like whole spheres [12].

Biebuyck et al. used selective dewetting of liquid prepolymers on a surface printed with self-assembled monolayers to fabricate arrayed microlenses made of an organic polymer [13]. Due to the low viscosity of the prepolymer required by this process, the microlenses fabricated using this method exhibited relatively small curvatures and thus short focal lengths [12].

Lu et al. described a self-assembly approach to fabricating patterned 2D arrays of microlenses with well-controlled lateral dimensions in the range of

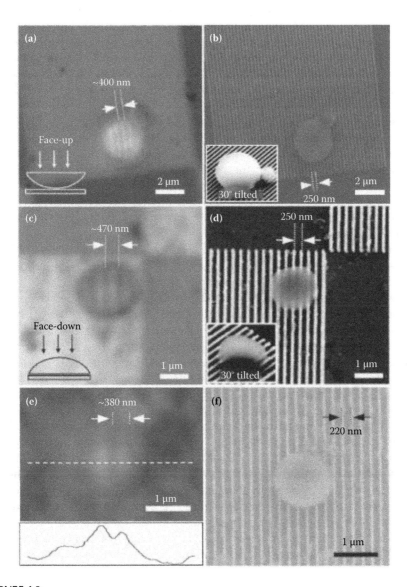

FIGURE 4.9
Optical microscope/SEM images of CHQ lenses on patterned substrates. (a) Optical microscope and (b) SEM image of a face-up lens placed on a glass substrate with palladium stripe patterns. (c) Optical microscope and (d) SEM image of a face-down lens. (e) Optical microscope and (f) SEM image of a face-up lens. Light intensity profile taken from the dotted line. The sub-diffraction-limit patterns cannot be resolved in conventional optical microscopy, but the magnifying effect through the lens allows the stripe patterns of 250/220nm spacing to be resolved. (*Source:* Lee, J.Y. et al. 2009. *Nature*, 460(7254), 498–501. With permission.)

1 to 10 μm [12]. Figure 4.10 illustrates the fabrication procedure. An aqueous dispersion of monodispersed polystyrene beads was confined within a packing cell composed of two glass substrates. The edge of this liquid slug was allowed to move slowly along the direction indicated by the arrow through a combination of evaporation and capillary flow of water. The 2D array of cylindrical holes previously defined in a PR film on the bottom substrate served as physical traps for retaining the liquid and positioning the polymer beads. As the edge of this liquid slug moved, the beads were dragged by capillary force across the surface of the bottom substrate and physically trapped by the holes.

The maximum number of polymer beads that could be held in each hole was governed mainly by geometry—specifically the ratios between the diameter D and height H of the holes and the diameter d of the polymer beads. This a very high-yield process; the self assembly success rates were as high as 95%. Because polystyrene is optically transparent down to the UV regime, these self-assembled polystyrene beads could directly function as patterned 2D arrays of spherical microlenses.

Figure 4.11 shows SEM images of the 2D microlens array. Figure 4.11a illustrates the as-prepared sample, with one polymer bead in each cylindrical hole. Figure 4.11b shows the same sample after annealing at 96°C for ~10 min, followed by removal of the PR traps. The spherical beads had been converted into mushroom-shapes with cylindrical bodies and hemispherical heads. Next, these mushroom-shaped microlenses could be further transformed into hemispheres by annealing again at 96°C for ~10 min, as shown in Figure 4.11c.

The fill factor of these microlens arrays could be controlled by changing the ratio between the volumes of the cylindrical holes and the polymer beads. Figure 4.11 is an SEM image of a sample fabricated with 5.7 μm wide polystyrene beads using a 2D array of cylindrical holes of about 2 μm in height and about 5 μm in diameter as the templates. Note the increase in the fill factor from ~57 to ~79% between Figures 4.11b and d with the variations in bead sizes.

Figure 4.12 compares the images obtained by projecting an object (the letter F) through the microlens array in three shapes: (a) spherical; (b) hemispherical; and (c) mushroom-shaped. Note that the spherical microlenses were unable to form clear focused images because of severe aberrations. In comparison, both hemispherical and mushroom-shaped microlenses formed sharper, better-focused images, although there was difference in the focal lengths between the hemispherical and mushroom-shaped microlenses.

4.6 Self Assembly of Microlens Arrays Using Global Dielectrophoretic Energy Wells

Huang et al. presented a method of using dielectrophoretic (DEP) energy wells to self assemble microballs to form a microlens array [7]. They

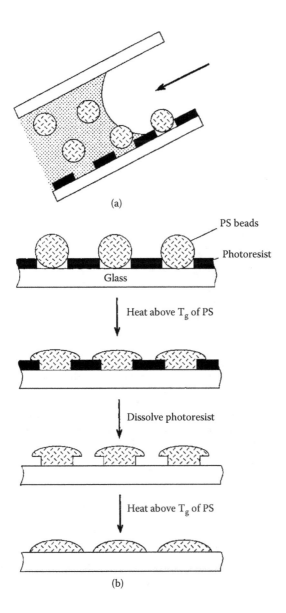

FIGURE 4.10
Experimental procedure used to fabricate two-dimensional arrays of polymeric microlenses on glass substrates. (a) Cross-sectional view of packing cell used to deliver monodispersed polystyrene beads into two-dimensional array of cylindrical holes patterned in thin film of PR spin coated on bottom glass substrate. (b) Fabrication of microlenses with hemispherical and mushroom-shaped profiles by annealing sample at temperatures above glass transition temperature of polystyrene (~93°C). The formation of a hemispherical shape was driven by the minimization of the surface free energy. (*Source:* Lu, Y., Y.D. Yin, and Y.N. Xia. 2001. *Advanced Materials*, 13(1), 34–37. With permission.)

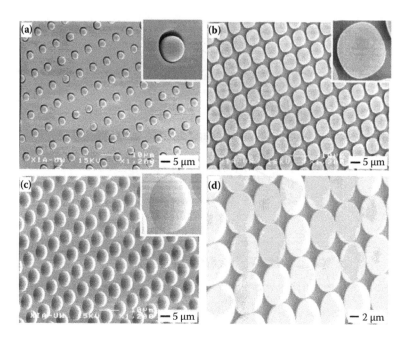

FIGURE 4.11

(a) SEM image of 2D array of 4.3 μm polystyrene beads trapped in a patterned array of cylindrical holes (2 μm in height and 5 μm in diameter) etched in a thin film of PR. (b) SEM image of 2D array of polystyrene beads after annealing at 96°C for ~10 min, followed by dissolution of the PR with ethanol. (c) SEM image of sample after it was annealed again at 96°C for ~10 min. In this second step of thermal annealing, the array of mushroom-shaped microlenses took on hemispherical form. (d) SEM image of 2D array of 5.7 μm polystyrene beads self-assembled in a patterned array of cylindrical holes (2 μm in height and 5 μm in diameter). Because the volume of each polymer bead was increased while the dimensions of the cylindrical hole remained unchanged, the hemispherical microlenses touched each other when the polymer beads were softened. (*Source:* Lu, Y., Y.D. Yin, and Y.N. Xia. 2001. *Advanced Materials*, 13(1), 34–37. With permission.)

generated DEP energy wells to position one electrically polarizable microball per energy well. The energy wells were formed under AC voltages on a spatially patterned dielectric material that distorted the uniform electric fields between two conductive parallel plates.

The DEP energy wells induced by the patterned dielectric were then applied to self assemble microlens arrays made from polystyrene (PS) microballs in water. The DEP force is determined by:

$$F_{DEP} = \frac{1}{4}\pi a^3 \varepsilon_m Re \left| K^* (\omega) \right| \nabla \left(E^2 \right) \tag{4.1}$$

where a is the microball diameter; ε_m is the permittivity of the surrounding medium; E is the strength of the electric field; $K^*(\omega)$ is the Clausius-Mossotti

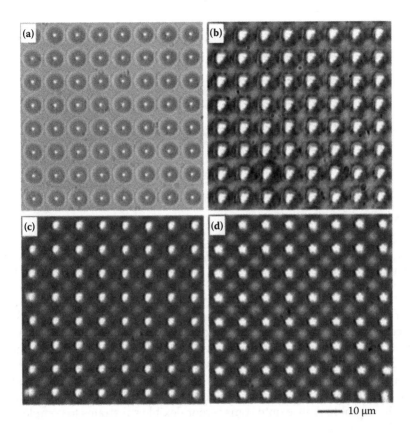

FIGURE 4.12
Images recorded with a transmission optical microscope by projecting object F through 2D array of polystyrene microlenses in three different shapes: (a) spherical, (b) hemispherical, and (c) mushroom-shaped. (d) Image of star-shaped object formed through 2D array of mushroom-shaped microlenses. (*Source:* Lu, Y., Y.D. Yin, and Y.N. Xia. 2001. *Advanced Materials*, 13(1), 34–37. With permission.)

(CM) factor (ranging from –1/2 to 1) at an AC frequency ω. The CM factor of the PS microballs in water was negative at AC signals of 30 kHz. The resulting negative DEP force pushed the PS microballs toward the side with weaker electric fields.

Figure 4.13 depicts the device used to self assemble microballs as microlens arrays. Figure 4.14 shows the DEP energy profile during the self assembly. First, an AC signal of 5 V was applied between two conductive indium tin oxide (ITO) layers. The dielectric material was a PR (AZ4620 in Figure 4.13) 5.5 μm in thickness. Photolithography was performed to define the DEP energy wells according to the desired pattern of the microlens array, as shown in Figure 4.14a.

Additional DEP energy gradients that were optically induced were also applied to compensate for array defects such as vacancies and excessive

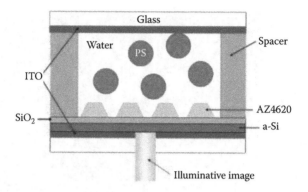

FIGURE 4.13
Device for self assembly of microballs for microlens arrays. Two ITO glasses are sandwiched by a 70 μm thick spacer. The bottom ITO glass is first coated with a layer of amorphous silicon followed by an insulating SiO_2 coating. A patterned PR is formed on the surface of the SiO_2 via photolithography. (*Source:* Huang, J.Y., Y.S. Lu, and J.A. Yeh. 2006. *Optics Express*, 14(22), 10779–10784. With permission.)

microballs. Optical intensity of 5 W/cm² from a liquid crystal (LC) display projector increased the conductivity of the amorphous silicon layer at the region of interest and consequently raised the DEP energy level as shown in Figure 4.14b. When the optical power swept across the array, excessive PS microballs were dislocated and relocated and vacancies were filled, as shown in Figure 4.14c. Finally, the PS microball arrays were transferred from the self-assembly device onto transparent flexible substrates to complete the microlens arrays.

This self-assembly method was able to produce microlenses on flexible substrates. Figure 4.15 shows SEM and optical microscope (OM) images of self-assembled microlens arrays on such flexible substrates. Figure 4.16a shows an OM image of the letter F on a transparency projected through a square microlens array. Figure 4.16b shows the interference pattern of the microlens array in a hexagonal array. The experimental setup used to obtain the images in a and b is shown in Figure 4.16c.

Focal length, numerical aperture (NA), image resolution, and uniformity in microlens arrays were also investigated. The initial focal length of the microball lenses was measured as 15.5 μm, corresponding to a NA of 0.80. The authors also measured the variation in the focal length to be lower than 0.8 μm, indicating good optical uniformity throughout the microlens array. The resolution of the microballs in the microlens array was 0.4 μm.

Due to the material properties of PS at temperatures over its glass transition temperature Tg of 90°C, PS microballs reflow partially or entirely and their geometries would thus change. The bottom semi-hemispheres of the PS microballs undergo significant shape change, but the top semi-hemispheres fortunately remain almost the same. The slight alteration of the top semi-hemispheres resulted in a slight change in the focal length and NA of the

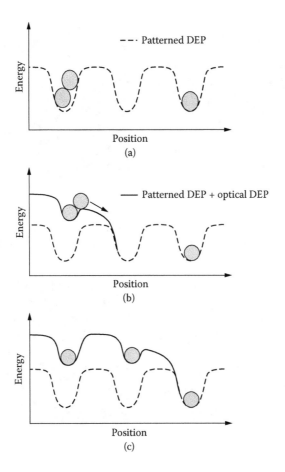

FIGURE 4.14
DEP energy over three adjacent energy wells. (a) Patterned DEP. (b) Illumination near first energy well. (c) Illumination swept to second energy well. The patterned DEP denotes energy produced by the patterned dielectric; the optical DEP denotes the energy induced by illumination. (*Source:* Huang, J.Y., Y.S. Lu, and J.A. Yeh. 2006. *Optics Express*, 14(22), 10779–10784. With permission.)

microlenses, as shown in Figure 4.17. Microballs heated at higher temperatures had longer focal lengths and lower NA values because of larger lateral elongation.

4.7 Microlens Arrays Fabricated from All-Liquid Techniques

Moench et al. explored three different liquid fabrication strategies to construct polymer microlens arrays on glass substrates [14]. First the glass substrates

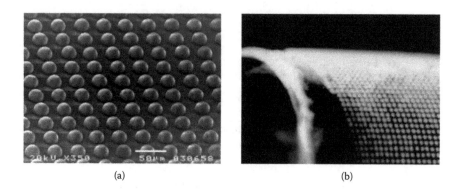

(a)　　　　　　　　　　　　　　　　　　(b)

FIGURE 4.15
(a) SEM image of microlens array on flexible substrate. (b) Optical micrograph of flexible microlens array curved at radius of 3 mm. (*Source:* Huang, J.Y., Y.S. Lu, and J.A. Yeh. 2006. *Optics Express*, 14(22), 10779–10784. With permission.)

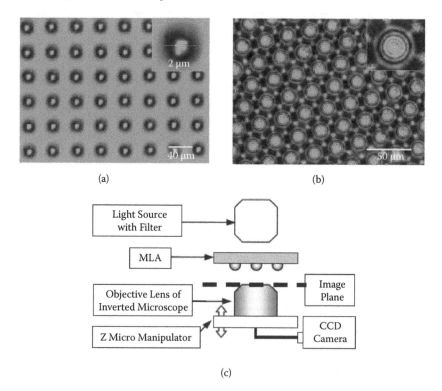

(a)　　　　　　　　　　　　　　　　　　(b)

(c)

FIGURE 4.16
Optical micrograph of projected images through the microlens array self-assembled from microball lenses 25 μm in diameter and numerical aperture of 0.80. (a) Image of lenses arranged in square pattern. (b) Interference pattern of lenses arranged in hexagonal pattern. (c) Experimental setup for imaging. (*Source:* Huang, J.Y., Y.S. Lu, and J.A. Yeh. 2006. *Optics Express*, 14(22), 10779–10784. With permission.)

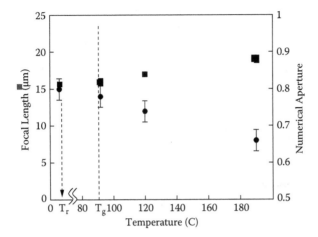

FIGURE 4.17
Change in focal length and numerical aperture for different post heat treatments. The heat treatment temperatures were 91, 120, and 190°C (all above the T_g of PS of 90°C) generated on a hot plate for 5 min. ■ = focal length. ● = numerical aperture. (*Source:* Huang, J.Y., Y.S. Lu, and J.A. Yeh. 2006. *Optics Express*, 14(22), 10779–10784. With permission.)

were patterned by microcontact printing using a suitable coupling agent. Next, a liquid monomer was selectively deposited on circular domains of the substrate as defined by the previous microcontact printing step. Three different methods were used for the deposition of the liquid monomer: repelling, dipping, and withdrawal. After subsequent cross linking of the monomer, the polymer microlenses were formed.

All three methods of depositing the liquid monomer involved dewetting. If dewetting occurs under water, the stronger interaction causes the water to repel the monomer from the hydrophilic domains and forces it to remain on the hydrophobic domains. This is the underlying principle for the repelling method [14] and the dipping method [13].

The two approaches are in fact quite similar. The difference is that in the repelling method, water is added to the monomer solution for the dewetting process, and in the dipping method, the sample with applied undiluted monomer is dipped into a water reservoir; accurate control of the dipping speed is achieved through a motorized stage. If the monomer film dewets in air, the monomer remains on the hydrophilic domains of the substrate. This is the case with the withdrawal method [15,16] in which the patterned substrate is completely immersed in the monomer and pulled out at a fixed angle and a controlled speed into ambient air.

After application of the monomer, subsequent polymerization can then be performed with UV irradiation in an inert gas (nitrogen) atmosphere. An example of a microlens fabricated by this technique is shown in Figure 4.18.

The focal length may be determined by a method described in Chapter 2. The surface profile of the lenses was first determined using scanning white

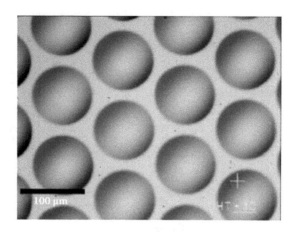

FIGURE 4.18
Hexagonal array of microlenses fabricated by withdrawal method to apply liquid monomer. The lenses consisted of a cross linked poly-alkyl-methacrylate network. Lens diameter = 90 µm. Focal length = 870 ± 30 µm. (*Source:* Moench, W. and H. Zappe. 2004. *Journal of Optics A,* 6(4), 330–337. With permission.)

light interferometry (SWLI). Then from the surface profile and refractive index of the microlens material, the focal length of the lenses was derived. For comparison, the authors also measured the focal length of the lenses using an optical microscope by focusing a collimated light at a wavelength of 543 nm that was transmitted through the lenses. Figure 4.19a shows the focal lengths of microlenses from 200 to 1300 µm. The measurements from both methods agreed well. Along with diameters of the microlenses, the *f* numbers were also calculated. They ranged from 4.0 to 13.2, as shown in Figure 4.19b.

4.8 Microlenses Produced by Direct Photo-Induced Cross Linking Polymerization

Croutxe-Barghorn et al. presented an approach using the self-developing characteristics of certain photopolymers to generate microlens arrays. The fabrication process takes advantage of the shrinkage of polymers—a mass transfer phenomenon due to gradients of chemical composition—and the bending of the surface when a photosensitive material is exposed to UV or visible light [17].

The fabrication process is shown in Figure 4.20. The material was a mixture of multifunctional acrylic monomers that could be polymerized with UV or visible light, depending on the photoinitiators incorporated into the polymer. The photopolymer mixture was coated onto glass slides and illuminated through an array of light spots generated by an argon ion laser emitting at 514.5 nm.

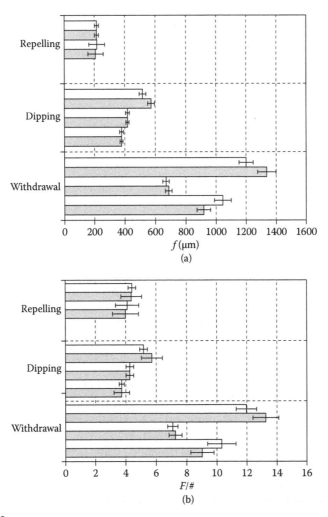

FIGURE 4.19
(a) Measured focal length of microlens arrays. (b) F numbers of microlens arrays. White bars indicate data from SWLI measurement; grey bars indicate optical microscope measurement. Each bar represents the measurement of 100 lenses in a 10 × 10 array. (*Source:* Moench, W. and H. Zappe. 2004. *Journal of Optics A*, 6(4), 330–337. With permission.)

The laser irradiation induced the formation of polymer cylinders in the bulk of the film. Acrylic monomers underwent a volume decrease of 10 to 15% when polymerized. The underlying reason was that the gradient of chemical composition resulting from the spatially controlled conversion of monomer into polymer immediately induced the flowing of monomer and photoinitiator molecules to the illuminated areas. Hollows created by volume shrinkage were almost immediately filled by reactive species.

Nevertheless, during this polymerization process, the surface of the film remained liquid. The reason was that the radical-induced polymerization

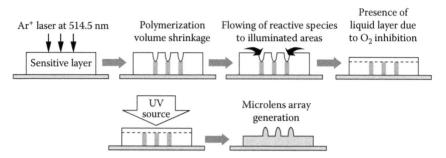

FIGURE 4.20
Microlens generation by photopolymerization. (*Source:* Croutxe-Barghorn, C., O. Soppera, and D.J. Lougnot. 2000. *Applied Surface Science*, 168(1–4), 89–91. With permission.)

was highly sensitive to oxygen (O_2) diffusion, and macroradicals generated through the chain reaction initiated by the reactive species could not escape atmospheric O_2 inhibition.

The total curing of the sensitive film was subsequently obtained by post-illuminating the entire surface with UV. At these wavelengths, the second photoinitiator incorporated in the sensitive mixture reacted and led to the formation of radicals that completed the polymerization. This homogeneous irradiation step induced monodimensional (in thickness) volume shrinkage of the sample. The illuminated areas that already consisted of 3D polymer cylinders with a thin layer of liquid monomer underwent less volume decrease so that the dark regions were still in the liquid form. As a result, areas polymerized with visible light emerged from the surface. Curing of the thin liquid layer and resulting surface tensions reshaped the tops of the polymerized cylinders into microlenses [17].

Figure 4.21 shows microlens arrays generated using this photopolymerization method. Figure 4.21a depicts the lenses generated by focusing the laser beam on the photosensitive layer. The microlenses in Figure 4.21b were obtained by illuminating the master array in Figure 4.21a with a laser beam of larger diameter and positioning the sensitive layer at the focal length of the lenses.

Figure 4.22 shows the relationship of the focal length of such microlens arrays to the parameters that determined the profile of the generated optical element. As shown, the shapes and focal lengths of the microlenses depend strongly on the illumination time due to polymerization kinetics including the numbers of radicals capable of initiating polymerization, fluidity of the photosensitive mixture, and mass transfers of reactive species [17].

4.9 Microlenses Formed by Thermally Reflowing Photoresists

Microlenses with fixed focal lengths made from thermally reflowed PRs emerged around 20 years ago [18–20] and have been widely studied and

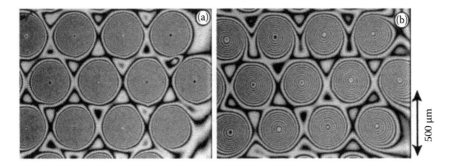

FIGURE 4.21
(a) Master microlens array generated with an argon ion laser (18 mW/cm^{-2}) focused to a 100 μm spot. Lens diameter = 400 μm. Height = 4.3 μm. (b) Microlens array. Lens diameter = 400 μm. Height = 2.5 μm. (*Source:* Croutxe-Barghorn, C., O. Soppera, and D.J. Lougnot. 2000. *Applied Surface Science*, 168(1–4), 89–91. With permission.)

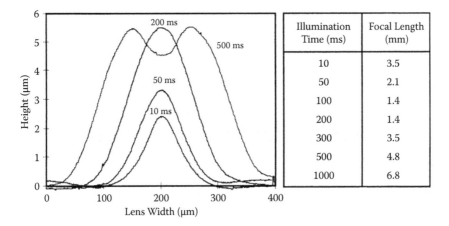

Illumination Time (ms)	Focal Length (mm)
10	3.5
50	2.1
100	1.4
200	1.4
300	3.5
500	4.8
1000	6.8

FIGURE 4.22
Relationship of shape, height, and focal length to irradiation time. The laser beam with intensity of 4.7 mW/cm^{-2} was focused on a 100 μm spot. (*Source:* Croutxe-Barghorn, C., O. Soppera, and D.J. Lougnot. 2000. *Applied Surface Science*, 168(1–4), 89–91. With permission.)

utilized [1,21–26]. This process uses only standard semiconductor fabrication equipment and processes (PR coating, photolithography, wet processing, etching, etc.) and allows the fabrication of large microlens arrays of excellent optical quality for wavelengths from the deep UV to the far IR.

The fabrication procedures are shown in Figure 4.23. First, a thin base layer of positive PR is spin coated on a substrate. Normally this base layer is less than 1 μm thick and is used to control the surface parameters of the substrate [18]. A polymerization bake is used to harden the PR. Next, a second layer (typically 1 to 100 μm thick) of positive PR is coated on top of the base layer, followed by standard exposure (Figure 4.23a) and developing (Figure 4.23b). An array of PR cylinders is obtained.

FIGURE 4.23
Fabrication of microlenses by the reflow or resist-melting method. (a) Photolithography. (b) Developing. (c) Melting of resist structure. (*Source:* Nussbaum, P. et al. 1997. *Pure and Applied Optics,* 6(6), 617–636. With permission.)

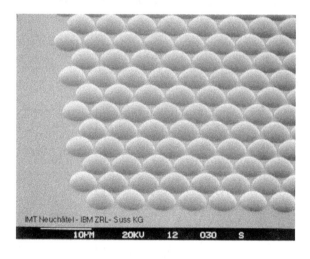

FIGURE 4.24
SEM image of microlens array fabricated by PR reflow method. (Ø = 5 µm, hexagonally densely packed). (*Source:* Nussbaum, P. et al. 1997. *Pure and Applied Optics,* 6(6), 617–636. With permission.)

The PR cylinders are melted at a temperature of 150 to 200°C on a hot plate or in an oven. During the melting procedure, the surface tension minimizes the surface area and thus rearranges the liquid masses. Ideally, if the PR melts completely, the masses are transported freely and spherical microlenses are formed. In reality, a complete melting of the PR is not always achievable, especially with large flat PR cylinders. For large PR volumes, the outer part of the liquid drop may have already cross linked (due to out-gassing of solvents) before the inner part is completely melted.

For different lens diameters, heights, and array types (packing densities, array sizes, substrate materials), all process parameters such as exposure energy, developing, prebaking, cooling, storing conditions, melting cycle, etc., must be optimized carefully [21]. Figure 4.24 is an SEM image of a typical microlens array fabricated by the PR reflow method.

To form gapless microlens arrays, Pan et al. developed a process based on PR reflow and metal electroplating to form array molds [22]. Figure 4.25 shows

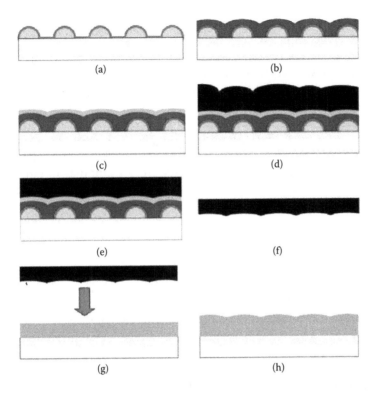

FIGURE 4.25
Flow chart of gapless microlens process. (a) Sputtering Ni thin film onto half-spherical PR structure as seed layer. (b) Creating primary master mold by electroplating Ni–Co mold. (c) Post-electroplating passivation on surface of primary master mold. (d) Second electroplating to fabricate secondary master mold. (e) Leveling surface by CMP. (f) Demolding to obtain master mold. (g) Fabricating microlens array by hot embossing. (h) Demolding and obtaining gapless triangular microlens array. (*Source:* Pan, C.T. and C.H. Su. 2007. *Sensors and Actuators A,* 134(2), 631–640. With permission.)

the reported fabrication process. First, the triangular microlens array was formed by thermally reflowing PR. Then nickel–cobalt (Ni-Co) electroplating technique was applied to transfer the PR pattern into a metal Ni-Co mold.

The metallization process required five steps. First, Ni thin film was sputtered on the surface of the triangular reflowed structure (Figure 4.25a). The Ni would then serve as the seed layer for the subsequent Ni-Co electroplating in the second step to form the Ni-Co mold (Figure 4.25b). The completed Ni-Co mold serves as the primary master mold.

Third, passivation was applied on the surface of the Ni-Co primary master mold (Figure 4.25c). This treatment involved sputtering of silver (Ag) thin film 250 nm in thickness on the surface of the primary master mold. Fourth, Ni-Co electroplating was applied again to deposit Ni and Co on the Ag thin film to obtain the inverse mold of the primary master. Since the adhesion between Ag and Ni-Co is poor, the two molds separated easily after the electroplating was completed.

The inverse (or secondary master) mold of the primary master mold (Figure 4.25d) was then obtained. Fifth, chemical mechanical polishing (CMP) was applied to flatten the mold (Figures 4.25e and f). Finally, the secondary master mold served as the master mold for a hot embossing process to transfer the desired structure onto a sheet of poly(methyl methacrylate) (PMMA; see Figure 4.25g). After demolding, PMMA-based microlens arrays were obtained (Figure 4.25h). Depending on the shapes and sizes on the mask, the shapes of the microlenses vary from triangular to circular [22]. Figure 4.26 shows the number of lenses of different shapes in the same area.

4.10 Microlens Arrays Fabricated with Polymer Jet Printing Technology

MacFarlane et al. used microjet (or inkjet) technology to fabricate microlens arrays [23]. Figure 4.27 shows the setup to generate microdroplets. The microjet system consisted of a piezoelectric ceramic with a microchannel. A nozzle exit was aligned at one end of the channel and a capillary tube intake connected to a reservoir was fitted at the other end. The reservoir was heated to lower the viscosity of the jetted polymer materials.

A pulse flexed the channel and forced a droplet through the aperture. This droplet traveled toward a substrate mounted on an XYZ micropositioner. As the molten polymer landed on the substrate, it cooled and solidified into a planoconvex shape, thus forming a microlens. The polymer could undergo a subsequent reflow, as in the PR reflow process. The speed of cooling and subsequent reflow were controlled by a substrate heater.

In this fabrication technique, typical microlens diameters ranged from 70 to 150 μm and the focal lengths were between 20 and 70 μm [23]. Jain et al. used this method to fabricate a single lens as a scanner; an SEM image is shown in Figure 4.28a [4]. Choo et al. also applied this technology to generate an addressable microlens array, as shown in Figure 4.28b, to improve the dynamic range of Shack-Hartmann sensors [5].

Dorrer et al. utilized a similar concept but used a dipping method to fabricate microlenses from a commercially available monomer–prepolymer mixture, Norland Optical Adhesive 74 (NOA74). They investigated the swelling of the microlenses in solvent vapors and determined the resulting changes in the focal length [24]. The fabrication of the microlenses is outlined in Figure 4.25.

Glass microslides were first coated with a 10 nm thick fluoropolymer layer in a dip coating step. The thin film was then structured with circular hydrophilic domains through photoablation. Short wavelength UV radiation was

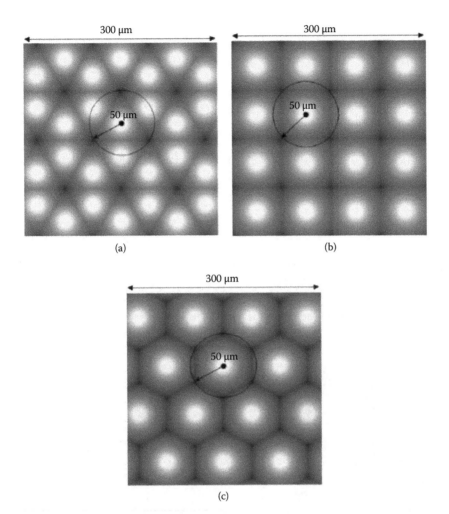

FIGURE 4.26
Number of lenses within an area of 300 μm × 300 μm. (a) Gapless triangular microlens array with 28 lenses. (b) Gapless square microlens array with 16 lenses. (c) Gapless hexagonal microlens array with 14 lenses. (*Source:* Pan, C.T. and C.H. Su. 2007. *Sensors and Actuators A,* 134(2), 631–640. With permission.)

used to "photovolatilize" the fluoropolymer in areas not protected by a photomask. The diameter of the hydrophilic circles was 400 to 1000 μm.

A standard needle printer was customized by attaching a 2 μL glass pipette to a print head. The needle printer first took up polymer solution by dipping into a reservoir. The amount of liquid that adhered to the needle depended on many parameters such as solution viscosity, contact angle of the solution on the needle, and velocity of the needle as it left the reservoir. Then the needle was moved to the substrate, coming down onto the hydrophilic spots from above, and was stopped 40 to 70 μm above the surface.

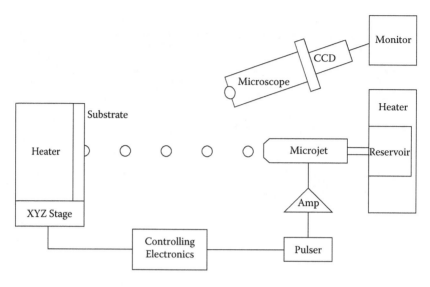

FIGURE 4.27
Microjet system for forming droplets on demand. The droplets are used to manufacture microlenses. (*Source:* MacFarlane, D.L. et al. 1994. *IEEE Photonics Technology Letters*, 6(9), 1112–1114. With permission.)

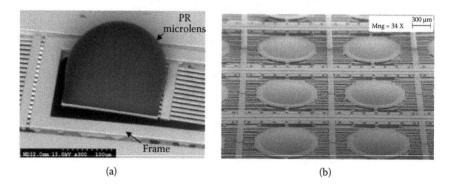

(a)   (b)

FIGURE 4.28
(a) SEM image of lens holder integrated with PR microlens as a scanner. (*Source:* Jain, A. and H.K. Xie. 2005. *IEEE Photonics Technology Letters*, 17(9), 1971–1973. With permission.) (b) SEM image of addressable microlens array for Shark-Hartmann sensors. (*Source:* Choo, H. and R.S. Muller. 2006. *Journal of Microelectromechanical Systems*, 15(6), 1555–1567. With permission.)

Although the needle did not touch the substrate, a drop of polymer solution clinging to the needle came into contact with the hydrophilic spots. A certain amount of polymer solution was thus transferred and spread onto the circular hydrophilic area. The substrate was then placed on a hot plate to evaporate any remaining solvent. Subsequent UV irradiation led to the formation of covalent bonds between the polymer chains, and a polymer network and resultant microlenses were formed.

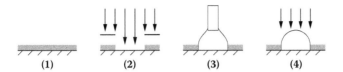

FIGURE 4.29
Fabricating microlenses with a dipping method. (1) Fluoropolymer thin film is deposited on the surface by dip coating. (2) In areas not protected with a photomask, polymer is decomposed by low wavelength UV radiation. (3) Polymer solution is transferred onto hydrophilic spots using a needle printer. (4) After evaporation of solvent, the polymer is cross linked in a UV flood exposure. (*Source:* Dorrer, C., O. Prucker, and J. Rühe. 2007. *Advanced Materials*, 19(3), 456. With permission.)

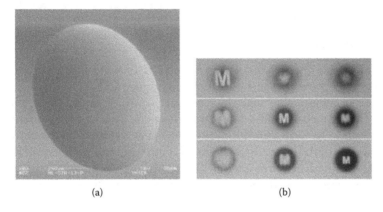

(a)   (b)

FIGURE 4.30
(a) SEM image of NOA74 microlens with diameter of 1000 μm. (b) M is imaged through 1000 μm microlenses with different focal lengths. Each row shows the same three lenses, but the *f* of the microscope has been systematically varied (down the columns). The M is clear only when the image is in focus. (*Source:* Dorrer, C., O. Prucker, and J. Rühe. 2007. *Advanced Materials*, 19(3), 456. With permission.)

An SEM image of a microlens is shown in Figure 4.30a. The focal length of these microlenses was determined by the process parameters. Figure 4.30b illustrates microlenses with different focal lengths (770, 1030, and 1740 μm, respectively).

4.11 Microlens Arrays Fabricated through Molding

Zeng et al. presented an approach to making polydimethylsiloxane (PDMS) microlens arrays [26] through liquid phase photopolymerization (LP³) [25] and molding. Figure 4.31 shows the basic configuration of the microlens array. The aperture array that generated the microlens array was defined by a single photomask. The apertures were connected through a microchannel.

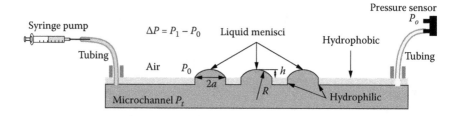

FIGURE 4.31

Basic principle of forming microlens arrays by molding process. The radii of curvature of liquid menisci and thus the focal lengths of the resultant microlenses are adjusted by controlling the pneumatic pressure in the microchannel during photopolymerization of the liquid menisci. The radius of the aperture is a; the height and the radius of curvature of the liquid menisci are h and R, respectively; and the pressures in the microchannel and in air are P_l and P_0, respectively. (*Source:* Zeng, X. and H. Jiang. 2008. *Journal of Microelectromechanical Systems*, 17(5), 1210–1217. With permission.)

One end of the microchannel was connected to a syringe pump and the other end to a pressure sensor, as shown in Figure 4.31.

A photopolymerizable water-based liquid was first flowed into the microchannel and pneumatically squeezed from the apertures by operating a syringe pump to form curved liquid menisci at the liquid–air interfaces at the top edges of the apertures. The radii of curvature of these liquid menisci were determined by three factors (1) the difference between the applied pneumatic pressure and the atmospheric pressure, (2) the surface tension, and (3) the gravity of the liquid. A specific radius of curvature and thus the focal length of the resultant microlenses could be achieved by tuning the applied pneumatic pressure P_l in the microchannel.

The sidewalls and bottom surfaces of the apertures were chemically treated to be hydrophilic. The top surfaces of the apertures were naturally hydrophobic based on material surface properties, thereby pinning the liquid–air interfaces along the hydrophobic–hydrophilic boundaries at the top edges of the apertures.

These liquid menisci were maintained in a certain convex shape through pneumatic control and throughout the subsequent photopolymerization of the polymer under UV irradiance, after which the polymer transfer molds were formed. Due to its excellent light transparency and mechanical flexibility, PDMS was used as the microlens material. To produce convex PDMS microlens arrays, two molding steps were needed to transfer the fabricated polymerized molding structures to PDMS. Figure 4.32 shows the process flow [26].

Figure 4.33 shows the optical image of a typical microlens array fabricated as described above. The SEM image, the side view image taken by a goniometer, and the magnified image of an object using the microlens array are also shown. The diameter of each microlens in the array was 1.8 mm.

Figure 4.34 plots the relationship of the focal lengths of the microlens arrays made from the above molding process with varying diameters of the

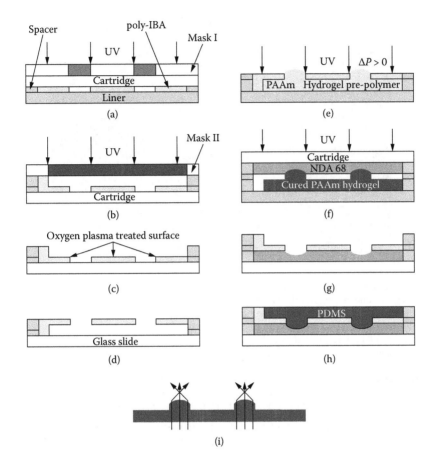

FIGURE 4.32
Fabrication of microlens array by molding. (a) Isobornyl acrylate (IBA)-based prepolymer mixture solution is flowed into a 250 μm thick cartridge well, and the aperture array is photopatterned using LP³. (b) Poly-IBA microchannel is photopatterned in a 350 μm thick cavity using LP³. (c) Sidewalls and top surfaces of poly-IBA plate are treated with oxygen plasma to form hydrophobic–hydrophilic pinning boundaries. (d) Cartridge plate is peeled off and the poly-IBA plate is flipped over and bonded onto a glass slide. (e) Liquid menisci of water-based polyacrylamide (PAAm) hydrogel prepolymer solution is formed under pneumatic pressure and polymerized under flood UV irradiance. (f) Nonshrinkable Norland Optical Adhesive (NOA) 68 is applied onto the polymerized structures of PAAm hydrogel and precured under flood UV exposure. (g) Cured PAAm hydrogel is dissolved in a chemical solution and precured NOA 68 is next fully cured under UV irradiance for 4 hours to form a mold. (h) PDMS prepolymer mixture is applied onto the fully cured NOA 68 mold and cured at room temperature. (i) PDMS microlens arrays are peeled from the NOA 68 mold to complete the fabrication. (*Source:* Zeng, X. and H. Jiang. 2008. *Journal of Microelectromechanical Systems*, 17(5), 1210–1217. With permission.)

apertures d and applied pneumatic pressure difference ΔP. As the diameter of the lens aperture d increased, the effect of the gravity increased and the curve deviated more than the result without the gravity.

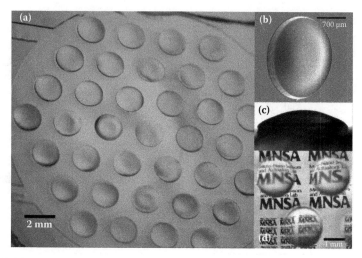

FIGURE 4.33

(a) Optical image of a microlens array, taken with a stereomicroscope from an oblique angle. (b) SEM image of one microlens taken from a 40°angle. (c) Side-view image of the same microlens taken with a goniometer. (d) Magnified image of an object using the microlens array. The film is under the microlens array, and the distance between them is 3 mm. The diameter of each microlens is 1.8 mm. Reprinted from [26] with permission from IEEE. (*Source:* Zeng, X. and H. Jiang. 2008. *Journal of Microelectromechanical Systems*, 17(5), 1210–1217. With permission.)

4.12 Microlens Arrays Fabricated by Hot Intrusion

Pan et al. introduced a microintrusion process developed and modified from a microembossing process to make plastic microlenses [27]. The process flow of the microintrusion process is illustrated in Figure 4.35. First, Ni mold inserts were made on a silicon substrate. A sheet of polycarbonate (PC) film 500 μm thick was placed underneath the mold insert as shown in Figure 4.35a. The film and insert were pressed firmly at a temperature above the glass transition temperature of the PC film (140 to 150°C).

When adequate pressure was applied, the plastic material deformed and intruded into the circular openings in the Ni mold as shown in Figure 4.35b. At the end of the process, the silicon substrate of the Ni mold insert was released (Figure 4.35c) for further processing or the plastic microlenses could be demolded, as shown in Figure 4.35d. Figure 4.36 shows SEM images of a microlens with the Ni mold insert and after the demolding process, respectively.

Figure 4.37 shows the characterization results of such microlenses fabricated by micro-hot intrusion. Images and relative intensities of a single-mode optical fiber with and without coupling with a microlens were obtained to show the function of the lens. A single mode optical fiber was aligned with a laser beam from a 623.8 nm He-Ne laser. The intensity distribution of the output from the fiber was converted from the image. As can be seen, focusing with the microlens led to higher intensity of the output light spot.

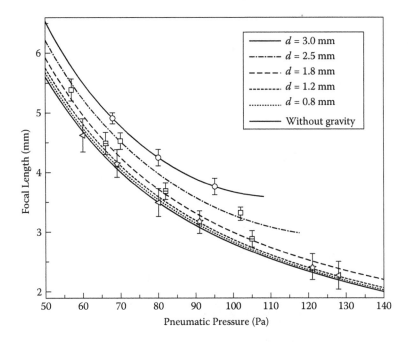

FIGURE 4.34
Focal length f of microlens arrays as function of pneumatic pressure ΔP and aperture diameter d. (*Source:* Zeng, X. and H. Jiang. 2008. *Journal of Microelectromechanical Systems*, 17(5), 1210–1217. With permission.)

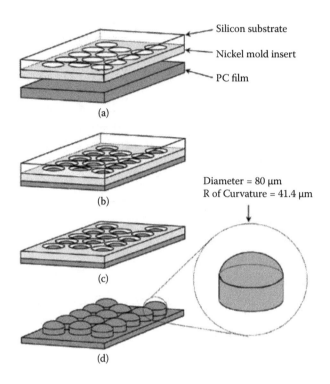

Silicon substrate

Nickel mold insert

PC film

(a)

(b)

Diameter = 80 μm
R of Curvature = 41.4 μm

(c)

(d)

FIGURE 4.35
Fabrication sequences of micro-hot intrusion. (a) Mold insert and polycarbonate film. (b) Micro-hot intrusion at elevated temperature and pressure. (c) Release of silicon substrate. (d) Release of plastic lenses. (*Source:* Pan, L.W., X. Shen, and L. Lin. 2004. *Journal of Microelectromechanical Systems*, 13(6), 1063–1071. With permission.)

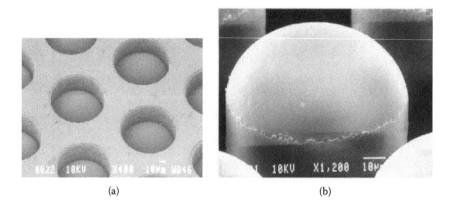

(a)

(b)

FIGURE 4.36
(a) SEM of fabricated microlens array with Ni mold insert. (b) SEM image of microlens after demolding. (*Source:* Pan, L.W., X. Shen, and L. Lin. 2004. *Journal of Microelectromechanical Systems*, 13(6), 1063–1071. With permission.)

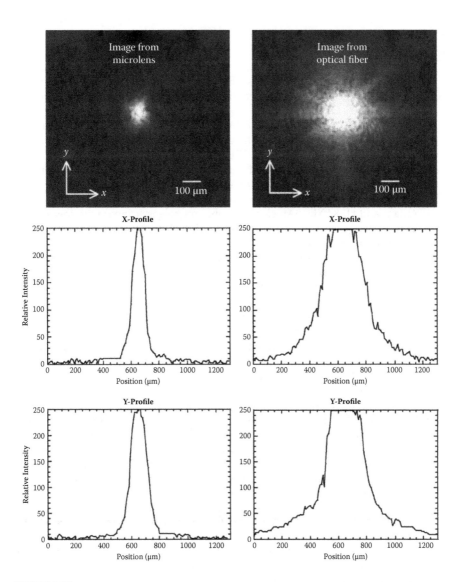

FIGURE 4.37
Images and relative intensities of an optical fiber with (left) and without (right) coupling with a microlens from an 80 μm diameter mold insert. (*Source:* Pan, L.W., X. Shen, and L. Lin. 2004. *Journal of Microelectromechanical Systems*, 13(6), 1063–1071. With permission.)

References

1. M. H. Wu and G. M. Whitesides, "Fabrication of diffractive and micro-optical elements using microlens projection lithography," *Advanced Materials*, vol. 14, pp. 1502–1506, Oct 2002.
2. M. H. Wu and G. M. Whitesides, "Fabrication of two-dimensional arrays of microlenses and their applications in photolithography," *Journal of Micromechanics and Microengineering*, vol. 12, pp. 747–758, Nov 2002.
3. R. Volkel, H. P. Herzig, P. Nussbaum, P. Blattner, R. Dandliker, E. Cullmann, and W. B. Hugle, "Microlens lithography and smart masks," *Microelectronic Engineering*, vol. 35, pp. 513–516, Feb 1997.
4. A. Jain and H. K. Xie, "An electrothermal microlens scanner with low-voltage large-vertical-displacement actuation," *IEEE Photonics Technology Letters*, vol. 17, pp. 1971–1973, Sep 2005.
5. H. Choo and R. S. Muller, "Addressable microlens array to impove dynamic range of Shack-Hartmann sensors," *Journal of Microelectromechanical Systems*, vol. 15, pp. 1555–1567, Dec 2006.
6. N. F. Borrelli, "Efficiency of microlens arrays for projection LCD," in *Electronic Components and Technology Conference, 1994. Proceedings., 44th*, 1994, pp. 338–345.
7. J. Y. Huang, Y. S. Lu, and J. A. Yeh, "Self-assembled high NA microlens arrays using global dielectricphoretic energy wells," *Optics Express*, vol. 14, pp. 10779–10784, Oct 2006.
8. H. C. Chen, S. W. Lee, and L. J. Chen, "Self-aligned nanolenses with multilayered Ge/SiO2 core/shell structures on Si (001)," *Advanced Materials*, vol. 19, pp. 222–226, Jan 2007.
9. M. Eisner and J. Schwider, "Transferring resist microlenses into silicon by reactive ion etching," *Optical Engineering*, vol. 35, pp. 2979–2982, Oct 1996.
10. J. Y. Lee, B. H. Hong, W. Y. Kim, S. K. Min, Y. Kim, M. V. Jouravlev, R. Bose, K. S. Kim, I. C. Hwang, L. J. Kaufman, C. W. Wong, P. Kim, and K. S. Kim, "Near-field focusing and magnification through self-assembled nanoscale spherical lenses," *Nature*, vol. 460, pp. 498–501, Jul 2009.
11. S. Hayashi, Y. Kumamoto, T. Suzuki, and T. Hirai, "Imaging by Polystyrene Latex-Particles," *Journal of Colloid and Interface Science*, vol. 144, pp. 538–547, Jul 1991.
12. Y. Lu, Y. D. Yin, and Y. N. Xia, "A self-assembly approach to the fabrication of patterned, two-dimensional arrays of microlenses of organic polymers," *Advanced Materials*, vol. 13, pp. 34–37, Jan 2001.
13. H. A. Biebuyck and G. M. Whitesides, "Self-Organization of Organic Liquids on Patterned Self-Assembled Monolayers of Alkanethiolates on Gold," *Langmuir*, vol. 10, pp. 2790–2793, Aug 1994.
14. W. Moench and H. Zappe, "Fabrication and testing of micro-lens arrays by all-liquid techniques," *Journal of Optics A: Pure and Applied Optics*, vol. 6, pp. 330–337, Apr 2004.
15. D. M. Hartmann, O. Kibar, and S. C. Esener, "Characterization of a polymer microlens fabricated by use of the hydrophobic effect," *Optics Letters*, vol. 25, pp. 975–977, Jul 2000.

16. D. M. Hartmann, O. Kibar, and S. C. Esener, "Optimization and theoretical modeling of polymer microlens arrays fabricated with the hydrophobic effect," *Applied Optics*, vol. 40, pp. 2736–2746, Jun 2001.

17. C. Croutxe-Barghorn, O. Soppera, and D. J. Lougnot, "Fabrication of microlenses by direct photo-induced crosslinking polymerization," *Applied Surface Science*, vol. 168, pp. 89–91, Dec 2000.

18. S. Haselbeck, H. Schreiber, J. Schwider, and N. Streibl, "Microlenses Fabricated by Melting a Photoresist on a Base Layer," *Optical Engineering*, vol. 32, pp. 1322–1324, Jun 1993.

19. Z. D. Popovic, R. A. Sprague, and G. A. N. Connell, "Technique for Monolithic Fabrication of Microlens Arrays," *Applied Optics*, vol. 27, pp. 1281–1284, Apr 1988.

20. D. Daly, R. F. Stevens, M. C. Hutley, and N. Davies, "The Manufacture of Microlenses by Melting Photoresist," *Measurement Science & Technology*, vol. 1, pp. 759–766, Aug 1990.

21. P. Nussbaum, R. Volke, H. P. Herzig, M. Eisner, and S. Haselbeck, "Design, fabrication and testing of microlens arrays for sensors and microsystems," *Pure and Applied Optics*, vol. 6, pp. 617–636, Nov 1997.

22. C. T. Pan and C. H. Su, "Fabrication of gapless triangular micro-lens array," *Sensors and Actuators A-Physical*, vol. 134, pp. 631–640, Mar 2007.

23. D. L. MacFarlane, V. Narayan, J. A. Tatum, W. R. Cox, T. Chen, and D. J. Hayes, "Microjet Fabrication of Microlens Arrays," *IEEE Photonics Technology Letters*, vol. 6, pp. 1112–1114, Sep 1994.

24. C. Dorrer, O. Prucker, and J. Rühe, "Swellable surface-attached polymer microlenses with tunable focal length," *Advanced Materials*, vol. 19, pp. 456–460, Feb 2007.

25. X. Zeng and H. Jiang, "Polydimethylsiloxane microlens arrays fabricated through liquid-phase photopolymerization and molding," *Journal of Microelectromechanical Systems*, vol. 17, pp. 1210–1217, Oct 2008.

26. A. K. Agarwal, S. S. Sridharamurthy, D. J. Beebe, and H. Jiang, "Programmable autonomous micromixers and micropumps," *Journal of Microelectromechanical Systems*, vol. 14, pp. 1409–1421, Dec 2005.

27. L.-W. Pan, X. Shen, and L. Lin, "Microplastic lens array fabricated by a hot intrusion process," *Journal of Microelectromechanical Systems*, vol. 13, pp. 1063–1071, Dec 2004.

5

Electrically Driven Tunable Microlenses

Unlike traditional glass lenses, focal lengths of microlenses can be tuned electrically or mechanically by various mechanisms. In this chapter we will describe a series of electrically tuned microlenses: (1) liquid lenses covered with a thin polymer film and driven by electrostatic forces; (2) lens arrays utilizing dielectrophoretic effect; (3) electrochemically activated liquid lenses; (4) tunable liquid lenses actuated by electrowetting; and (5) liquid crystal (LC) lenses.

5.1 Introduction

Miniaturizing optical systems has been an active field of endeavor in recent years. The most important components in all optical systems are their lenses. In many cases, changing the focal distance of a system involves adjusting the whole system. However, such on-the-fly adjustments are often hard to realize for miniaturized systems. Developing miniature structures for tunable focus microlenses with variable focal lengths has attracted great interest. Many methods such as electrical fields and mechanical forces have been investigated. In this chapter, we will discuss electrically driven tunable microlenses.

5.2 Liquid Crystal Microlenses

5.2.1 Initial Designs

The use of nematic liquid crystals (LCs) for making tunable focus microlenses was described as early as 1979 by Sato. He showed that LC cells shaped into planoconvex or planoconcave lenses could be controlled electrically to form microlenses [1]. In 1984, Kowel et al. used a simple parallel electrode structure in a uniform thickness LC cell. Each electrode was subjected to different voltages to demonstrate the optical focusing ability of LC microlenses [2,3].

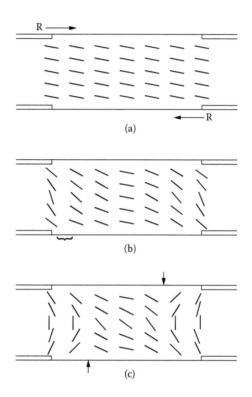

FIGURE 5.1

Liquid crystal molecular orientation model at various voltages. (a) No voltage. (b) Low voltage. (c) High voltage. R = rubbing direction. (*Source:* Nose, T., S. Masuda, and S. Sato. 1992. Japanese *Journal of Applied Physics Part 1*, 31(5), 1643–1646. With permission.)

Figure 5.1 shows a molecular orientation model in a cross section of LC microlens along a diameter of the hole. Figure 5.1a shows the molecular orientation model when no external voltage is applied. Note that the molecules are homogeneously aligned with a small pre-tilt angle after a surface rubbing treatment. When a relatively low voltage is applied, the LC molecules around the edge of the hole are forced to tilt according to the pre-tilt direction. No reversely tilted regions leading to a disclination line are observed, as shown in Figure 5.1b.

A molecular orientation model with a high voltage applied appears in Figure 5.1c. Since the intensity of the electric field decreases steeply upon approaching the center of the hole, the molecules near the center of the hole are not aligned along the electric field even at a relatively high voltage [4]. It is now clear that at this high voltage the orientation of the LC molecules becomes non-uniform and is distributed from the edge toward the center of the hole.

Since the LC is birefringent (refractive index depends on polarization and propagation direction of light), the non-uniform orientation of the molecules would lead to patterns observable under crossed polarizers. Figure 5.2

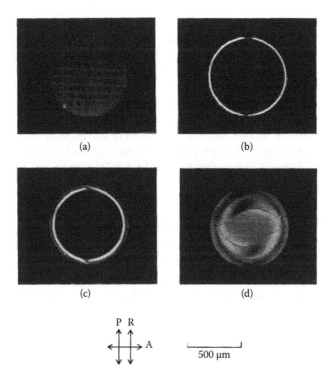

FIGURE 5.2
Transmitted light images observed using a polarizing microscope with crossed polarizers. The applied voltage levels were (a) 0, (b) 3, (c) 7, and (d) 70 V, respectively. The diameter of the hole pattern was 700 μm and cell thickness was 50 μm. (*Source:* Nose, T., S. Masuda, and S. Sato. 1992. *Japanese Journal of Applied Physics Part 1*, 31(5), 1643–1646. With permission.)

shows transmitted light images taken under a polarizing microscope with such crossed polarizers at various voltage levels. One polarization direction of incident light is parallel to the rubbing direction, and a tungsten lamp is used as a light source. The transmitted light image is uniformly dark without voltage, as shown in Figure 5.2a because all molecules are aligned roughly in one polarization direction and thus filtered out by the orthogonal polarizer.

When voltage is applied, LC molecules are forced to re-orientate and tilt from the surface of the substrate according to the intensity distribution of the electric field within the rubbing direction. However, the intensity of the electric field near the edge of the hole is so high that the molecules tend to twist along the axially symmetric distribution of the non-uniform electric field.

Figure 5.2b shows that a bright annular area appears because of birefringence and the twisted molecular orientation since the light now has components in both polarization directions. The annular area does not become as wide around the voltage level where good lens properties appear; hence, the linearly polarized incident light experiences a lens property without changes in polarization states.

As voltage increases, two tiny disclination lines appear near the edge of the hole pattern along the rubbing direction as shown in Figure 5.2c. They indicate the boundaries of the reversely twisted areas. These phenomena are the same as those of the hybrid aligned microlens with a LC of negative dielectric anisotropy. These two disclination lines may appear around the lower and upper substrates, respectively. When a very high voltage is applied (exceeding ~40 V), two disclination lines are observed in the direction perpendicular to the rubbing direction (Figure 5.2d), and the LC cell no longer exhibits good lens properties. These disclination lines are caused by a reverse tilting of the molecules, and each line suddenly appears with increasing voltage.

5.2.2 Improved Electrode Structures and Designs

Based on the molecular orientational tilting of LCs, various structures have been suggested to vary voltages onto LCs and thus change optical properties [5,6]. Choi et al. fabricated a two-dimensional LC microlens array using a surface relief structure of ultraviolet (UV)-curable polymer [5]. The microlens array cell was made with indium tin oxide (ITO) glass substrates. One substrate had a specific surface relief structure of the photopolymer used and the other had only the ITO layer.

To develop a surface relief structure, the UV-curable polymer was spin coated on the substrates at a rate of 4000 rpm for 30 sec. The resultant thickness of the spin coated polymer film was about 3 μm. After a polymer film formed on the glass substrate, spatially modulated UV light was irradiated onto the film through a photomask. The photomask was designed to generate the intensity modulation of the UV light to form a pattern of microlens arrays. The irradiation led to the diffusion of monomers in the polymer composites from the unexposed to exposed regions. The result was the formation of a lens-shaped surface relief structure.

The alignment layer was then prepared from a polyvinyl alcohol (PVA) solution (1 wt.%). In this situation, the average pre-tilt angle of the LC molecules is almost negligible on the alignment layer. The alignment layer coated over the surface relief structure determines the optical axis of the LC when a rubbing process is involved. Two cases of LC alignment on the surface relief structure (with and without the alignment layer) were examined [5]. The LC was filled into the sandwiched substrates by capillary action. Figure 5.3 shows the LC microlens structure and two operating states.

Riza et al. fabricated a three-terminal device for optical beam forming [6]. Figure 5.4 shows the top view of a thin film resistor-biased LC cylindrical lens, with electrical biasing resistors on the left and right sides of the active area. A spherical or elliptical lens can be formed by the appropriate use of a cascade of two such LC cylindrical lenses with orthogonal lens axes.

A thin film of transparent n+ amorphous silicon 5.33 nm thick with an area approximately 1 mm × 1 mm, was deposited onto a glass substrate. Next,

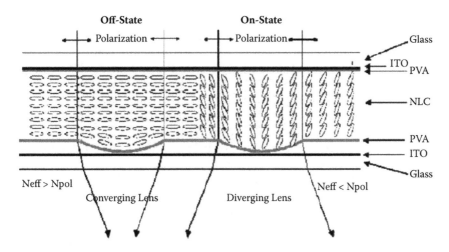

FIGURE 5.3
Liquid crystal microlens structure and two operating states. The crystal was aligned homogeneously on both surfaces. (*Source:* Choi, Y., J.H. Park, J.H. Kim et al. 2003. *Optical Materials,* 21(1–3), 643–646. With permission.)

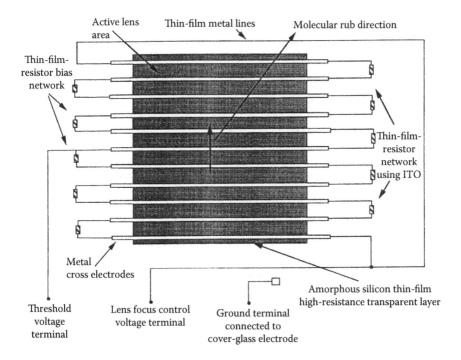

FIGURE 5.4
Top view of thin film resistor network-biased liquid crystal lens design. (*Source:* Riza, N.A. and M.C. Dejule. 1994. *Optics Letters,* 19(14), 1013–1015. Wth permission.)

highly conductive cross electrodes of molybdenum metal were evaporated over the n+ amorphous silicon layer. The electrodes were 2.5 μm wide with 12 μm pitch and ~1 mm long. The electrodes were opaque, so only light coming through the gap could be transmitted through the n+ film. An ITO thin film resistor network with a resistance of 450 Ω/square was built on the left and right sides of the lens array to bias the electrodes to simulate a cylindrical lens.

A constant-amplitude ac (e.g., 1 kHz square-wave) signal was applied to the top and bottom terminals (or electrodes) of the LC device (Figure 5.4) with the center electrode biased near the LC threshold value (~1 V) for molecular activation. The top cover-glass electrode was grounded. A piecewise quadratically varying index perturbation was generated for the cylindrical LC lens by the electrode structure, using a single driver or voltage level. By varying the voltage level the focal length of this lens could then be changed.

Because the voltage on the electrodes is designed to change as the square of the distance from the center of the lens device, a square change in the LC index of refraction occurs if the LC is operated in the linear portion of its characteristic curve. Figure 5.5 illustrates LC molecules in their rotated positions as results of the different voltage levels generated by the resistor bias network in the device. n_e represents the extraordinary index of refraction of the LC material, n_o is the ordinary index of refraction, and $\Delta n = n_e - n_o > 0$ is the positive optical anisotropy of the birefringence of the LC mixture.

To generate a short-focal-length lens, a relative optical shift between the center of the lens and the edges of the lens should be maximized. The device shown was 9 μm thick and filled with E63 LC (Merck KGaA, Darmstadt, Germany) material ($\Delta n = 5.227$ at 589 nm). The molecules had a parallel-rub configuration. The directions of the molecular rub director are shown in Figure 5.4 and Figure 5.5.

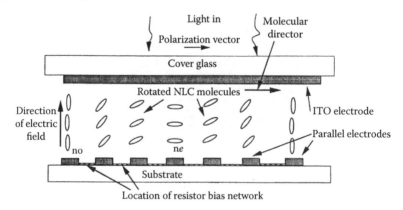

FIGURE 5.5
Side view of device showing various positions of electrically controlled liquid crystal molecules used to generate the quadratic optical phase shifts required to form a variable focal length cylindrical lens. (*Source:* Riza, N.A. and M.C. Dejule. 1994. *Optics Letters,* 19(14), 1013–1015. With permission.)

5.2.3 Liquid Crystal Microlens Arrays Using Dielectrophoretic (DEP) Effect

Cheng et al. demonstrated an LC droplet-based microlens driven by DEP force [7]. LC is a high dielectric medium and thus may be potentially actuated by DEP forces. Based on the relative permittivity $\varepsilon_{//}= 16.5$ and conductivity $\sigma = 5 \times 10^{-12}$ $(\Omega \cdot m)^{-1}$ for the LC MDA2625 (Merck), the factor $\sigma/(\omega \varepsilon_b \varepsilon_0)$ is calculated to be 5×10^{-6} at an operation frequency of 1 kHz. By definition of dielectric material of $\sigma/(\omega \varepsilon_b \varepsilon_0) \ll 1$, LC thus behaves as a dielectric medium. Using this DEP mechanism Cheng's group demonstrated the adjustment of the focal length of the liquid crystal droplet lens.

Figure 5.6 illustrates the operation schematics of the LC droplet lens. Nonuniform electric fields generated by concentric electrodes yield DEP forces exerted on the droplet. The ITO electrodes (50 µm in width and 50 µm in spacing) were fabricated on an ITO glass wafer and coated with Teflon coating. The Teflon functioned as a hydrophobic layer, increasing the contact angle of the liquid crystal droplet to the surface. The DEP mechanism used to deform the droplet can be described using the Kelvin polarization volume force density f as described in

$$\vec{f} = \frac{1}{2}\varepsilon_0 \left(\varepsilon_{//} - 1 \right) \nabla \left| \vec{E} \right|^2 \tag{5.1}$$

where $\varepsilon_{//}$ and ε_0 are the permittivities of LC and free space, respectively. E denotes the electric field intensity.

Figure 5.7 shows the images of a letter F at different focal lengths of a 2 µL LC droplet lens in the isotropic phase. The images were captured by a CCD

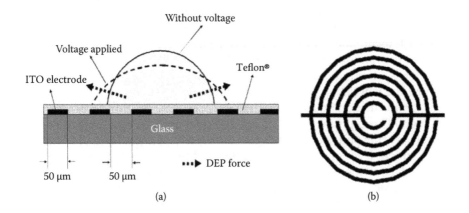

(a)   (b)

FIGURE 5.6
(a) Mechanism of a deformable liquid crystal droplet lens (not to scale). (b) Design of concentric ITO electrode of 50 µm width and 50 µm spacing. (*Source:* Cheng, C.C., C.A. Chang, and J.A. Yeh. 2006. *Optics Express*, 14(9), 4101–4106. With permission.)

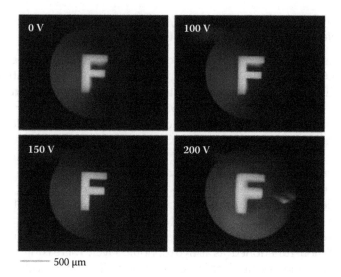

— 500 µm

FIGURE 5.7
Tuning focal length of a liquid crystal droplet lens under various applied voltages. The temperature was kept at 30°C. (*Source:* Cheng, C.C., C.A. Chang, and J.A. Yeh. 2006. *Optics Express,* 14(9), 4101–4106. With permission.)

sensor on a microscope. The images appeared in focus at 150 V and became blurry when the voltages were tuned away from 150 V. As the voltage decreased to 100 V or lower, the LC droplet lens had shorter focal length.

5.2.4 Liquid Crystal Microlens Arrays Using Patterned Polymer Networks

As described previously, polymer patterning techniques can be used to define LC microlenses. However, it is relatively difficult to control lens shapes with such technique, and the lens surface thus formed is rather rough as a result of the random phase separation between the LC and the prepolymer. Therefore, obtaining microlens arrays with high optical performance while keeping the fabrication process simple remains a challenging task.

Ren et al. demonstrated a method of preparing microlens arrays with polymer–LC composite materials [8]. The polymer was first molded to bear microlens array patterns through a lamination technique. Instead of coating an alignment layer on the concave surfaces of the polymer microlens arrays, Ren used a polymer network to stabilize the LC alignment. The introduction of the polymer network into the LC bulk was found to restrain the disclination of the LC during the focus change and greatly improve the switching speed because of the strong anchoring force of the polymer networks.

In comparison with other tunable microlenses, this fabrication method can obtain an ideal spherical shape and glazed surface for each microlens. Consequently, imaging quality and optical efficiency were much improved.

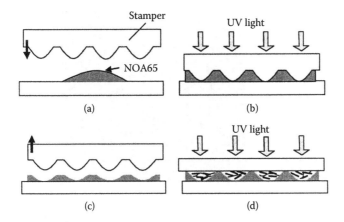

FIGURE 5.8
Procedures for fabricating polymer network liquid crystal microlens arrays. (*Source:* Ren, H.W., Y.H. Fan, and S.T. Wu. 2004. *Optics Letters*, 29(14), 1608–1615. With permission.)

Figure 5.8 depicts the procedure for fabricating such LC microlens arrays. The first step was to pour NOA65 prepolymer onto an ITO glass substrate. A planar glass-based convex microlens array (Isuzu Glass, Japan) was used as a stamper to laminate the NOA65 (Figure 5.8a). The thickness and the diameter of each microlens were 45 and 450 mm, respectively. The second step was to cure the laminated NOA65 with UV light (10 mW/cm² for 10 min), as shown in Figure 5.8b.

After UV exposure, the third step was to remove the stamper. At this stage, the solid planoconcave microlens arrays were formed on the bottom substrate as in Figure 5.8c. When the sample was heated from the stamper side, the stamper could be easily peeled off without damaging the concave polymer microlens. The final step was to inject the LC–monomer mixture into the recessed portions of the concave polymer microlenses and to seal them with a top ITO glass substrate (Figure 5.8d).

To obtain homogeneous LC alignment, the inner surface of the top substrate was coated with a polyimide layer and rubbed in one direction. The LC mixture was heated above its clear point and then slowly cooled to room temperature. The sealed cell was then exposed to UV light at 10 mW/cm² intensity for 60 min to form the needed polymer networks. The UV curing was conducted at room temperature.

Figure 5.9a illustrates the focused spot patterns produced by such microlens arrays when no voltage was applied. Arrays of uniform bright spots generated by positive lenses were observed. The light-intensity profile of spot A was measured at different voltages, V = 0, 40, 60 V$_{rms}$, respectively (Figure 5.9b). At V = 0, the peak intensity was the strongest. As the applied voltage increased, the LC directors in the polymer cavities reoriented toward the electric field and the curvature of the refractive index profile gradually

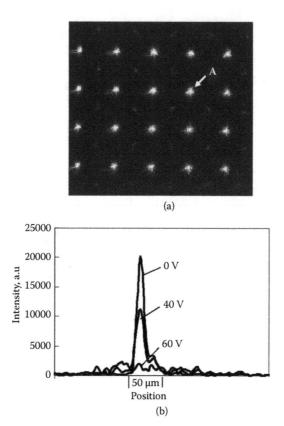

FIGURE 5.9
(a) Arrays of light spots produced by 450 μm diameter microlens array with 480 μm pitch. (b) Measured spot intensity profiles for spot A at V = 0, 40, and 60 V_{rms}. (*Source:* Ren, H.W., Y.H. Fan, and S.T. Wu. 2004. *Optics Letters,* 29(14), 1608–1615. With permission.)

flattened. As a result, the focal length of the microlens arrays increased and the measured peak intensity at the CCD focal plane decreased. When the voltage was further increased to 60 V_{rms}, the lenses gradually lost focusing capability and the resultant spot intensity decreased dramatically. The ordinary refractive index of the LC matched that of the NOA65 prepolymer, and thus the curvature of the refractive index gradient was nearly flattened.

5.2.5 Fresnel Lens Using Nanoscale Liquid Crystals

Ren et al. reported the fabrication of a Fresnel lens using a photomask with zone plate patterns [9]. The LC monomer in the zones cured with more UV intensity would lead to smaller LC droplets. Conversely, the zones with weaker UV exposure would result in larger nanoscale droplets. When a uniform voltage was applied to the zone plate, the refractive index (or phase) of the

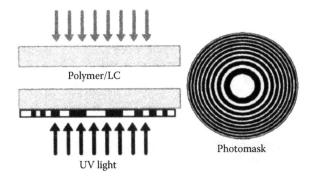

FIGURE 5.10
Left: Method for fabricating Fresnel lens. Right: Photomask showing Fresnel zone structure. (*Source:* Ren, H.W., Y.H. Fan, and S.T. Wu. 2003. *Applied Physics Letters,* 2003. 83(8): 1515-1517. With permission.)

zones with larger LC droplets changed so that the beam diffraction efficiency and focusing behaviors of the lens could be adjusted. Since the nanoscale LC droplets in the polymer matrix were randomly distributed, the resultant lens was polarization-independent and exhibited a fast switching time.

Figure 5.10 illustrates the fabrication of the Fresnel lens. The key element is the photomask with transparent odd zones and opaque even zones. The photomask was produced by etching a chromium oxide layer using electron beam lithography. The radius r_1 of the innermost zone was 5.5 mm and the radius of the nth zone (r_n) was given by $r_n^2 = nr_1^2$, where n is the zone number. The Fresnel zone plate consisted of 80 zones in a 1 cm aperture.

The odd zones of the prepared sample were highly transparent when the voltage was off. The even zones appeared slightly bluish, which implies that the formed LC droplet was comparable to the wavelength of blue light. The Fresnel zone plate was inspected using a polarizing optical microscope where the sample was set between two crossed polarizers. At $V = 0$, the birefringence colors of the Fresnel zones could be observed.

In a binary phase Fresnel lens, the focal length f is related to the innermost zone radius r_1 as $f = r_1^2/\lambda$, where λ is the wavelength of the incident beam. The primary focal length of the lens was ~50 cm (for $\lambda = 633$ nm). Due to the higher order Fourier components, a Fresnel zone lens has multiple foci at f, $f/3$, $f/5$, etc. However, most of the incident light diffracts into the primary focus. The theoretical diffraction efficiency of the primary focus for the binary-phase Fresnel lens is 41%.

To characterize the light focusing properties of the fabricated Fresnel lens, Ren's group measured the image quality, as shown in Figure 5.11. The collimated He-Ne laser beam had a diameter of ~1 cm and filled the entire zone plate. A CCD camera was set 50 cm from the Fresnel lens. A black cardboard with a transparent letter R was placed between lens L_2 and the sample. The CCD camera was set ~25 cm behind the sample.

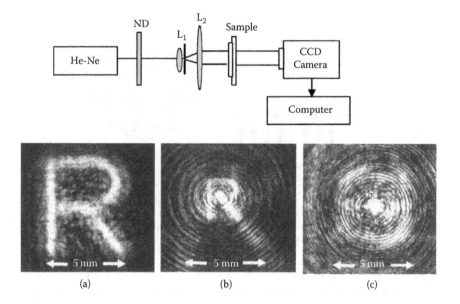

FIGURE 5.11
(a) Experimental setup for studying the Fresnel lens, (b) Imaging and focusing properties of the Fresnel lens recorded by a CCD camera. (a) Without sample, (b) CCD at 25 cm from sample, and (c) CCD at focal point. (Reprinted from [9] with permission from *American Institute of Physics*.)

The figure shows setup and operation with and without the sample. When the sample was absent, no focusing effect occurred. After the LC Fresnel lens was in position, a clear but smaller image was obtained, although some circular noise arose from diffraction. When the CCD camera was moved to the focal point, a small spot appeared in the center (Figure 5.11c).

Li et al. applied LC microlenses for adaptive ophthalmic lenses [10]. A Mach-Zehnder interferometer operating at 543.5 nm was used to measure wavefront quality immediately behind the lens and determine the focal length. The technique is based on interference between the spherical wave after the lens and a reference plane wave.

Figure 5.12a depicts an example of the measured interferograms. Five interferograms were taken with a phase shift between each interferogram and a wrapped phase map could be generated. The unwrapped phase map representing the actual optical path difference profile generated by the diffractive lens appears in Figure 5.12b. A good spherical wave was obtained with very few higher order aberrations as indicated by an rms wavefront error of only 5.0889λ.

The focal length was 55.855 cm. The focal length at 555 nm was calculated to be 49.80 cm, corresponding to a focusing power of 2.008 diopters. Diopter is another way of measuring the optical power of a lens. It is equal to the

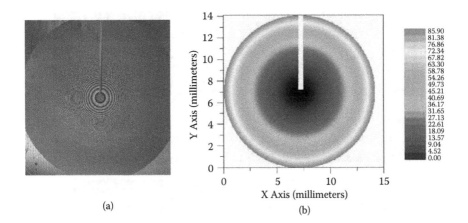

FIGURE 5.12
(a) Interferogram obtained with Mach-Zehnder interferometry of 15 mm, four-level, two-diopter lens. (b) Unwrapped phase map for 14 mm aperture. (*Source:* Li, G.Q., P. Valley, M.S. Giridhar et al. 2006. *Applied Physics Letters*, 89(14). With permission.)

reciprocal of the focal length measured in meters (1 diopter = 1 meter^{-1}). The focused spot size was also measured and found to be 47.9 μm, close to the diffraction-limited spot size of 45.1 μm. The response and decay times of this lens were measured at 180 and 120 msec, respectively. In addition, by changing the slope of the applied voltages to each zone, a negative 2-diopter focusing power could be obtained with the same lens, and the diffraction efficiency was the same as that for the positive 2-diopter case.

Li's group further demonstrated that by using nematic LCs, two LC lenses were set in orthogonal directions to operate for randomly polarized light. A white light source illuminated the object placed at a typical reading distance of approximately 30 cm. When the diffractive lens was not activated, the image was blurred, as shown in Figure 5.13a. However, when the diffractive lens was turned on, the object was imaged clearly (Figure 5.13b). The on and off states of the lens hence allow near and distance vision, respectively.

5.3 Liquid Microlenses Encapsulated with Polymer Thin Film Driven by Electrostatic Forces

Binh-Khiem et al. investigated a method to change the volume of the encapsulated liquid by electrostatic force [11]. The authors chemically deposited parylene directly on non-volatile liquids under low pressure conditions. The liquid surface was covered with a flexible thin polymer film and the liquid droplets had both shape and surface flexibility. Furthermore, the deposited

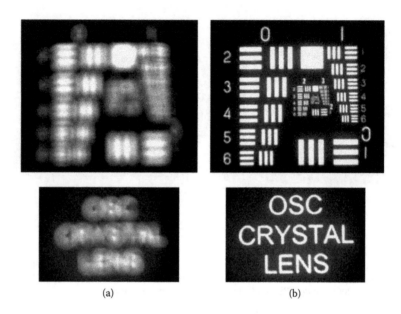

(a)   (b)

FIGURE 5.13

Imaging using two-diopter electroactive diffractive lens with model eye. The object is placed at reading distance (~30 cm). (a) The image is severely out of focus in the model eye when the diffractive lens is off. (b) When the diffractive lens is activated, the object is imaged clearly. (*Source:* Li, G.Q., P. Valley, M.S. Giridhar et al. 2006. *Applied Physics Letters,* 89(14). With permission.)

parylene film was smooth and transparent enough for optical use. The shapes of the encapsulated liquid droplets could be altered by electrostatic interactions. Figure 5.14a shows the structure of such a liquid lens packaged in a parylene thin film. The lens liquid was encapsulated in a parylene film less than 1 μm thick. A 5 nm thick gold layer deposited on the parylene film served as the upper electrode. The liquid microlenses were fabricated on a commercially available glass wafer sputtered with an ITO layer patterned to create the lower electrode. A liquophobic layer of CYTOP, an amorphous fluorocarbon polymer, was spin coated onto the substrate. Next, an aluminum layer was deposited and patterned so that circular domains of CYTOP were exposed. The substrate was treated with oxygen plasma to etch away CYTOP on the exposed domains to reveal the liquophilic glass surface.

After the aluminum layer was completely removed, the substrate surface had liquophilic circular domains in which glass surface surrounded by liquophobic areas, where CYTOP still covered the surface, was exposed. Liquid droplets were deposited, by dipping or spin coating onto the substrate surface. These liquophobic and liquophilic patterns force the liquid to form droplets of the desired shape and size on the substrate surface. After depositing parylene onto the droplet surface, the upper electrode, a 5 nm layer of gold (sufficiently thin to be transparent) was deposited on the parylene film. The total transmittivity of the lenses was about 62% at a wavelength of 650 nm.

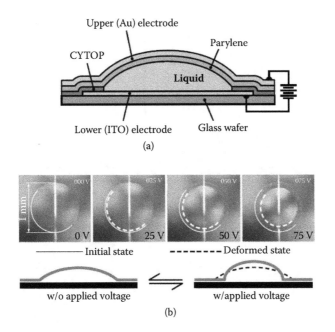

FIGURE 5.14
Conceptual structure and deformation of liquid lens packaged in parylene thin film. (a) Structure. (b) Deformation of the lens shape upon applying voltage. (*Source:* Binh-Khiem, N., K. Matsumoto, and I. Shimoyama. 2008. *Applied Physics Letters,* 93(12), 124105. With permission.)

When a voltage was applied, the electrodes accumulated a charge that then induced an electrostatic force that pulled the two electrodes toward each other. As shown in Figure 5.14b, the pulling force was mainly induced around its circular perimeter because of the short distance from the dome to the lower electrode, and thus became stronger. The upper electrode started to be attracted and approached the lower electrode from this perimeter.

Since the liquid volume inside was constant, any deformation from the initial spherical shape resulted in expansion of the parylene film in order to increase the total surface area. The height of the droplet increased while the radius of its circular base decreased until the electrostatic force balanced the elastic force from the parylene film caused by deformation.

Such deformation also caused the lens surface curvature to increase, which in turn decreased the focal length. The lens surface remained spherical throughout the deformation. When the applied voltage was removed, the droplet returned to its initial shape due to the elasticity of the parylene film.

A decrease in focal length to 20% of its initial value (3.7 mm) was measured when 150 V current was applied to a liquid lens 1 mm in diameter, 60 μm high, and packaged by 1 μm parylene C; see Figure 5.15b. The 5 mm diameter lenses could be switched by a rectangular voltage of 5 Hz with a delay shorter than 5.03 sec, Smaller lenses could be switched by voltages of higher frequencies. The lens actuation showed good repeatability with stable hysteresis.

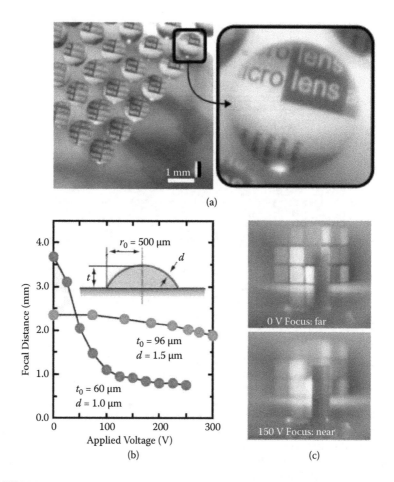

FIGURE 5.15

Varifocal liquid lenses. (a) A varifocal array with lens diameter of 1 mm. (b) Relationships of focal distance and applied voltage in fabricated 1 mm diameter lenses. (c) Changing focus with lens 10 mm in diameter. (*Source:* Binh-Khiem, N., K. Matsumoto, and I. Shimoyama. 2008. *Applied Physics Letters,* 93(12), 124105. With permission.)

5.4 Tunable Focus Liquid Microlenses Using Dielectrophoretic Effect

We previously described an LC droplet-based microlens developed by Cheng et al. [7]. The focal length was tuned using the dielectric force described in Section 5.2.3. Cheng further extended the design and introduced new liquids to a packaged liquid lens actuated by a dielectric force [12].

Figure 5.16a depicts the configuration of the new type of liquid lens actuated by dielectric force. The liquid lens consisted of a 15 µL liquid droplet with

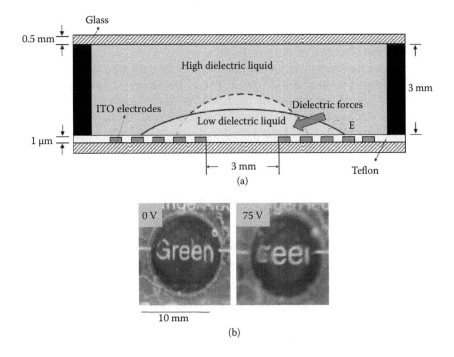

FIGURE 5.16
(a) Dielectric liquid lens. The droplet shrinks to a new state (dashed line) due to the dielectric force. (b) Captured images of an actuated liquid lens at the rest state (left) and at 75 volts (right). (*Source*: Cheng, C.C. and J.A. Yeh. 2007. *Optics Express*, 15(12), 7140–7145. With permission.)

a low dielectric constant and a sealing liquid with a high dielectric constant. The bottom diameter of the droplet was 7 mm when no voltage was applied. The two liquids were injected inside a 3 mm thick polymethyl methacrylate (PMMA) chamber formed and sealed between two ITO glass substrates. The concentric ITO electrodes on the bottom glass substrate were coated with 1 μm thick Teflon to reduce friction between the droplet and the glass substrate. The width and spacing of the ITO electrodes were both 50 μm.

An interesting property of the sealing liquid is that its mass density was adjusted to match that of the droplet to minimize the gravitational effect because the effect may induce non-uniform deformation of the droplet profile and cause optical aberrations. As the voltage was applied, a dielectric force was exerted onto the droplet due to the difference in the dielectric constants of the two liquids. The dielectric force shrank the droplet, increasing the contact angle of the droplet on the glass surface and shortening the focal length of the liquid lens. The dielectric force induced can be described by

$$\vec{f} = -\frac{1}{2}\varepsilon_0 \nabla \left[(\varepsilon_1 - \varepsilon_2) \left| \vec{E} \right|^2 \right] \tag{5.2}$$

where ε_0 is the permittivity of free space; ε_1 and ε_2 are dielectric constants of the sealing liquid and droplet, respectively. E denotes the electric field intensity across the interface of the two liquids.

The images in Figure 5.16b show the word *green* captured using the liquid lens at the rest state and at 75 V, respectively. The images were captured at a distance of 15 mm from the actuated liquid lens. At the rest state, the liquid lens had a long focal length due to the smaller intrinsic contact angle of the droplet. At 75 V, the shortened focal length rendered a magnified virtual image.

A similar mechanism was utilized to form liquid microlens arrays by Ren's group [13]. To form the liquid microlens array, two immiscible dielectric liquids were used. The first designated L-1 had a relatively low dielectric constant and high refractive index. The second (L-2) had a higher dielectric constant but lower refractive index.

Figure 5.17 illustrates the fabrication procedures and operation of the microlens array. Six fabrication steps appear in Figure 5.17a: (1) coating the electrode on the surface of a glass substrate, (2) etching the electrode with a pattern consisting of an array of holes, (3) coating a dielectric layer on the electrode surface, (4) coating L-1 on the dielectric layer, (5) dropping L-2 on the surface of L-1, and (6) laminating the two liquids using an ITO coated glass substrate. Care must be taken to avoid aggregation of L-1 when step 4 is finished; to this end, steps 5 and 6 should follow immediately.

Based on the figure, the formed L-1 layer could be flat, rough, or break into tiny droplets, depending on the surface tensions of the liquids. As soon as the cell was sealed, a suitable voltage was applied immediately (Figure 5.17b). Due to the hole array electrode, both liquids experienced a centrosymmetrical

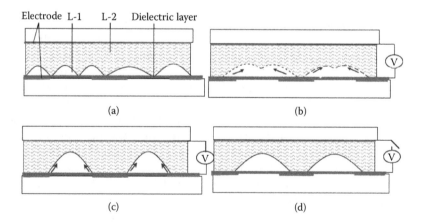

FIGURE 5.17
Side view of droplet array forming processes for liquid microlens array. (a) Two liquid layers. (b) Droplet forming process. (c) Stable state with voltage applied. (d) Relaxed state without voltage applied. (*Source:* Ren, H. and S.T. Wu. 2008. *Optics Express*, 16(4), 2646–2652. With permission.)

gradient electric field. Near the holes, a dielectric force was generated on the contact surfaces of the two liquids. The dielectric force can be expressed as

$$\vec{f} = \frac{1}{2}\varepsilon_0 \nabla \left[(\varepsilon_1 - \varepsilon_2) |\vec{E}|^2 \right] - \frac{1}{2} |\vec{E}|^2 \nabla (\varepsilon_1 - \varepsilon_2) \tag{5.3}$$

where ε_0, ε_1 and ε_2 are the permittivities of free space for L-1, and L-2, respectively, and E denotes the electric field across the cell gap. The second term is either negligible or zero because ε_1 and ε_2 are both scalars. Therefore, the electric field gradient plays the key role in generating the dielectric force. The net force is zero when $\varepsilon_1 = \varepsilon_2$. It should be also noted that the force changes its sign depending on whether the dielectric constant of the droplet or L-2 is larger.

A strong enough dielectric force could force L-1 and L-2 to redistribute. L-1 could be broken up into many small droplets and each droplet would be attracted by its nearby region with no electrodes (as the arrows show). The regions with the hole array electrode functioned as traps. If two or more L-1 droplets fell off the same trap, they would merge to form a single bigger droplet (Figure 5.17b)

L-2 would be squeezed into the high electric field region and surround the isolated L-1 droplets. After the redistribution of the two liquids, L-1 would exist as tiny droplets and occupy the weak electric field regions, as shown in Figure 5.17c. Due to the impact of the dielectric force, the droplets would be in the contraction state. When the voltage was removed, the droplets would spread around the electrode holes due to the potential energy balance (Figure 5.17d). The adjacent droplets would remain isolated because of the resistance of L-2. The distribution of L-1 was stable on the substrate surface because the following condition was satisfied

$$\gamma_{w,d} > \gamma_{o,w} + \gamma_{o,d} \tag{5.4}$$

where γ is the interfacial tension and the subscripts w, o, and d represent the water, oil, and dielectric layer, respectively. When the cell was activated again by voltage, the droplet surface returned to the shape in Figure 5.17c. The change in the droplet profile from Figure 5.17d to Figure 5.17c amounts to the tuning of the focal length of the microlens. Because the isolated liquid droplet was so small, theoretical analysis was used to determine that the shape of these ultrasmall droplets was mainly spherical.

In a relaxed state, the droplet would form a contact angle θ on the bottom substrate surface (Figure 5.17d). Since the shape of the droplet is axially symmetric, the droplet would function as a lens. In this state, the focal length of the droplet can be expressed as:

$$f^3 = \frac{3V}{\pi (1 - \cos\theta) \left(2 - (\cos\theta)^2 - \cos\theta \right) (n_1 - n_2)^3} \tag{5.5}$$

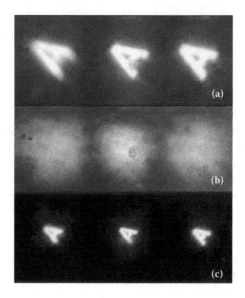

FIGURE 5.18
Imaging properties of microlens array at (a) $V = 0$, (b) $V = 60$ V, and (c) refocused state. (*Source:* Ren, H. and S.T. Wu. 2008. *Optics Express*, 16(4), 2646–2652. With permission.)

where V is the volume of the droplet, and n_1 and n_2 are the refractive indices of L-1 and L-2, respectively. When a voltage was applied to the electrode again, the induced dielectric force pushed the droplet and thus increased the contact angle. The focal length of the microlens will vary correspondingly.

The image quality of this microlens array is shown in Figure 5.18. A transparent object with a letter A was placed under the lens array. In the off state, a clear image array was first obtained by adjusting the distance between the lens cell and the object (Figure 5.18a). In contrast to the original object, the A image was inverted. As the applied voltage increased, the images became blurry instantly due to defocusing. Figure 5.18b shows the completely blurry image at 60 V (rms). Then the cell position was adjusted while the voltage was still applied to the cell. In Figure 5.18c, the image again appears clear.

The focal length of the single liquid lens was measured by a laser with a wavelength of 532 nm and a beam scanner; the reported values are shown in Figure 5.19 [13]. The focal length was determined for advancing actuation based on the minimum spot size resolved along the optic axis. The measured advancing focal lengths were compared with the paraxial approximation based on the advancing contact angles. The measurement results and paraxial approximation were in good agreement. When the voltage increased from 0 to 200 V, the focal length of the liquid lens varied from 34 to 12 mm.

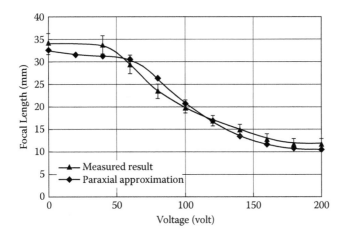

FIGURE 5.19
Focal lengths of dielectric liquid lens versus applied voltages. Triangles and diamonds indicate measurement results and paraxial approximations, respectively. (*Source:* Cheng, C.C. and J.A. Yeh. 2007. *Optics Express*, 15(12), 7140–7145. With permission.)

5.5 Tunable Focus Liquid Microlenses Using Electrowetting

5.5.1 Electrowetting

Electrowetting has become one of the most widely used tools for manipulating liquids by surface tension. Electrocapillarity, the basis of modern electrowetting, was first described in detail in 1875 by Gabriel Lippmann [14]. This ingenious physicist found that the capillary depression of mercury in contact with electrolyte solutions could be varied by applying a voltage between the mercury and the electrolyte. He subsequently formulated a theory of the electrocapillary effect and also developed several applications including a very sensitive electrometer and a motor.

This concept was recently revisited. In the early 1900s, Berge introduced the idea of using a thin insulating layer to separate the conductive liquid from the metallic electrode to eliminate the problem of electrolysis. This is the concept now known as electrowetting on a dielectric (EWOD) [15].

In electrowetting, one generically deals with droplets of partially wetting liquids on planar solid substrates, as shown in Figure 5.20. In most applications of interest, the droplets are aqueous salt solutions with a typical size of ~1 mm or smaller. The ambient medium can be air or an immiscible liquid; oil is often utilized. Under these conditions, the Bond number $B_0 = \sqrt{g\Delta\rho R^2/\sigma_{lv}}$ that measures the strength of gravity with respect to surface tension is smaller than unity. Therefore we can neglect gravity in most cases.

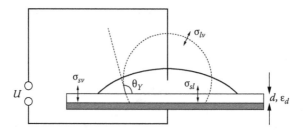

FIGURE 5.20
Generic electrowetting setup. Partially wetting liquid droplet at zero voltage (dashed line) and at high voltage (solid line). *Source:* Mugele, F. and J.C. Baret, 2005, *Journal of Physics: Condensed Matter*, 17(28), R705. With permission.)

In the absence of external electric fields, the behavior of the droplets is determined by surface tension alone. The free energy F of a droplet is a function of the droplet shape and expressed as the sum of the areas A_i of the interfaces between the three phases involved: the solid substrate (s), the liquid droplet (l), and the ambient phase assumed to be a vapor (v) for simplicity, weighted by the respective interfacial energies σ_i, (σ_{sv} for solid–vapor, σ_{sl} for solid–liquid, and σ_{lv} for liquid–vapor):

$$F = F_{if} = \sum_i A_i \sigma_i - \lambda V \tag{5.6}$$

Here, λ is a Lagrangian variable present to enforce the constant volume constraint and equals the pressure drop Δp across the liquid–vapor interface. Minimization of variation aids in maintaining the well-known conditions that any equilibrium liquid morphology must satisfy. These conditions have been described as the Young-Laplace equations discussed in Chapter 2. We can further break the procedure into two equations. The first is the Laplace equation stating that Δp is a constant, independent of the position on the interface:

$$\Delta p = \sigma_{lv}\left(\frac{1}{r_1} + \frac{1}{r_2}\right) = \sigma_{lv}k \tag{5.7}$$

Here, r_1 and r_2 are the two principal radii of curvature of the surface and κ is the constant mean curvature. For homogeneous substrates, this means that droplets will adopt the shape of a spherical cap in mechanical equilibrium. The second condition is given by Young's equation

$$\cos\theta_Y = \frac{\sigma_{sv} - \sigma_{sl}}{\sigma_{lv}} \tag{5.8}$$

that relates Young's equilibrium contact angle θ_Y to the interfacial energies. Alternatively, the interfacial energies σ_i can also be interpreted as

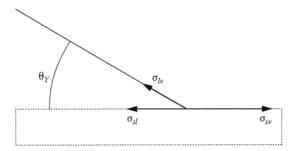

FIGURE 5.21
Force balance at contact line (for θ_Y approximately 30 degrees). *Source:* Mugele, F. and J.C. Baret, 2005, *Journal of Physics: Condensed Matter*, 17(28), R705. With permission.)

interfacial tensions, i.e., as forces pulling on the three-phase contact line. The equivalency between the two expressions has been described in Chapter 2. The physical meaning is sketched in Figure 5.21.

When a voltage is applied onto an electrolyte droplet on an electrode surface, as shown in Figure 5.20, the contact angle between the droplet and the surface follows the applied voltage.

$$\cos\theta = \cos\theta_Y + \frac{\varepsilon_0\varepsilon_1}{2d\sigma_{lv}}\left(V - V_0\right) \qquad (5.9)$$

where θ_Y and θ are the contact angles before and after applying the voltage, ε_0 and ε_1 are dielectric constants of the thin film, d is the thickness of the film, and V and V_0 are the initial and applied voltages, respectively. The contact angle thus decreases rapidly upon the application of a voltage. Figure 5.25, Figure 5.27, and Figure 5.32 (they appear in Section 5.5.3) show the relationship between contact angle change and applied voltage on the droplet.

5.5.2 First Liquid Lens Utilizing Electrowetting

One of the first proofs of concept for liquid lenses utilizing electrowetting was reported by Gorman et al. [16]. A drop of neat hexadecane thiol (HDT, ~1 pL) was placed on a bare gold surface under 100 mM aqueous sodium perchlorate electrolyte solution, as shown in Figure 5.22. The drop spread on the gold surface held at 0 V (in reference to a silver electrode) underwater and formed a hydrophobic monolayer under the drop. The advancing contact angle of this drop reached a value of 37 ± 2 degrees.

Switching the potential of the gold to −5.7 V versus the silver wire pseudoreference caused the electrode sorption of the hydrophobic self-assembled monolayer (SAM) and the retraction of the drop. The receding contact angle of this drop was 128 ± 2 degrees. As the drop retracted on the surface, water wetted the resulting charged gold surface. Retraction of the drop also reduced the interfacial area between the drop and the water as shown in the figure.

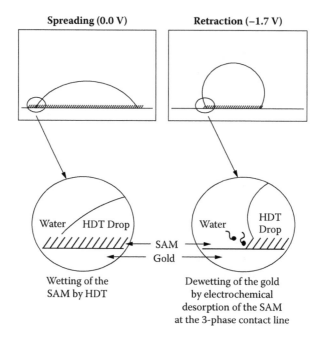

FIGURE 5.22

Behavior of drop of HDT on gold surface at 0 V (left) and at –5.7 V (right) compared with silver wire pseudoreference in 100 *mM* sodium perchlorate electrolyte solution. At 0 V, the HDT reacted with the gold surface and formed a monolayer wet by the HDT liquid. At –5.7 V, the monolayer reductively desorbed and the drop of HDT retracted. (*Source:* Gorman, C.B., H.A. Biebuyck, and G.M. Whitesides. 1995. *Langmuir,* 11(6), 2242–2246. With permission.)

A drop of HDT (~50 pL) on transparent gold may behave as a planar convex lens; see Figure 5.23. When the drop was on the surface at –5.7 V, it focused light that passed through it. Astigmatism of this lens was presumably minimal because the interfacial tension between the water and HDT ensured constant curvature of the lens and the interior of the lens was free of stress and thus exhibited few aberrations. A drop on gold held at 0 V also acted as a lens, but with a longer focal length. The drop could be perturbed electrochemically to bring an image repeatedly into or out of focus (>10 cycles). The response time of this lens was 25 msec. The relatively fast response of the electrowetting-based lens is one of its key advantages.

5.5.3 Tunable Liquid Microlens Utilizing Electrowetting

Electrowetting-based liquid lenses developed further after the landmark work by Gorman et al. Berge et al. reported an optical lens using a water droplet in 2000 [17]. Figure 5.24 illustrates the principle of the operation of the lens. A cell contained two immiscible liquids (one insulating and non-polar; the second was a conductive water solution). The liquids were transparent.

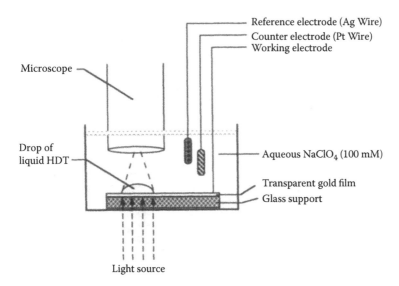

FIGURE 5.23
Procedure for electrochemical control of shape of HDT drop. The solution could be translated so that the microscope objective focused on the drop or the image focused by the drop. (*Source:* Gorman, C.B., H.A. Biebuyck, and G.M. Whitesides. 1995. *Langmuir*, 11(6), 2242–2246. With permission.)

They had different refractive indices with the same density so that gravity did not deform the liquid–liquid interface that remained spherical no matter what the cell orientation was.

The insulating liquid took the shape of a drop in contact with a thin insulating window. The surface of the window was hydrophobic, so that the insulating liquid naturally dwelled on it. A transparent (counter) electrode was deposited on the external side of the window. Application of voltage between the counter-electrode and the conducting liquid favored the wettability of the surface by the liquid. This deformed the interface and thus changed the focal length.

Figure 5.25 shows the variation of the reciprocal of the focal length in diopters as a function of applied voltage. The power of the lens was relatively stable below 90 V and started to rise when the applied voltage exceeded 90 V. Then a large focal length variation was observed. It should be noted that although the applied voltages seem very high for electrowetting-based lenses, the actual electrical power dissipated in the system was very small because power was consumed only during the switching and tuning of the lens. In other words, the current utilized by the operation of the lens is quite small. In this example, the dissipated power was a few milliwatts at 250 V.

Krupenkin et al. reported a planar electrode system for electrowetting based liquid lens [18]. They did not impregnate the substrate surface with a lubricant. Instead, they attempted to separate the droplet from the substrate

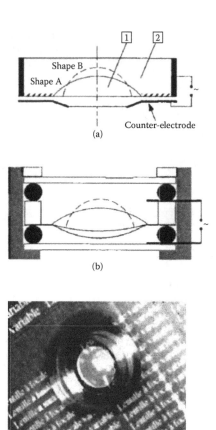

FIGURE 5.24

(a) Liquid lens using electrowetting (drawing not to scale). The cell was filled with water. A drop of an insulating non-polar liquid was deposited on the bottom wall constructed of a transparent insulating material. The central disc on the bottom wall surface was hydrophobic so it would trap the drop. The outer zone was hydrophilic. The optical axis was shown as a short long-dashed line. (b) Variable-focus liquid lens. The cell was sealed by o-rings. The central part diameter was 5 mm. (c) Photograph of one sample. Diameter of the stainless steel outer case was ~12 mm and the oil drop appears white at the center of the case. (*Source:* Berge, B. and J. Peseux. 2005. *European Physical Journal E*, 3, 159–163. With permission.)

surface by creating conditions under which a "lubricating" liquid would penetrate spontaneously under the droplet. This lubricating liquid was to be immiscible with the primary liquid forming the droplet. A thin layer of such a lubricant' would prevent the droplet from directly interacting with the contaminants and inhomogeneities of the solid surface, thus completely eliminating contact angle hysteresis and the stick–slip behavior [19]. Their tunable lens is shown in Figure 5.26.

To provide an electrical connection to the droplet without distorting its shape, a hole was made in the dielectric material so that the center of the droplet was

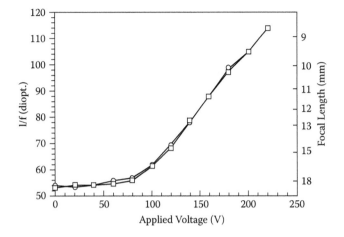

FIGURE 5.25
Focal length in diopters of variable focus lens (6 mm diameter) filled with α-chloronaphthalene as insulating liquid and a Na_2SO_4 solution in water as a function of voltage. The two curves corresponding to increasing and decreasing applied voltage are superimposed. (*Source:* Berge, B. and J. Peseux. 2005. *European Physical Journal E*, 3, 159–163. With permission.)

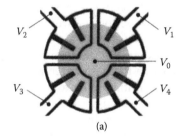

(a)

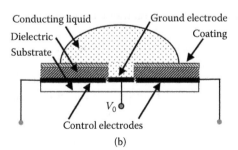

(b)

FIGURE 5.26
Tunable liquid microlens. (a) Electrode design. Black lines indicate etched areas on indium tin oxide (ITO). Applied voltages are indicated as V_0 through V_4. Lightly shaded area represents approximate droplet position. (b) Device cross section. Voltage applied to ground electrode is indicated as V_5. (*Source:* Krupenkin, T., S. Yang, and P. Mach. 2003. *Applied Physics Letters*, 82(3), 316–318. With permission.)

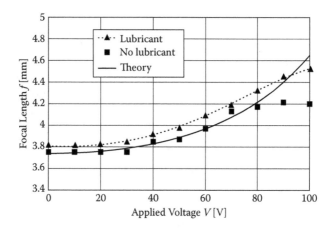

FIGURE 5.27

Microlens focal length versus applied voltage. Filled squares indicate data obtained without lubricating liquid. Filled triangles indicate data obtained with lubricating liquid. Dashed line is a guide for the eyes. Solid curve corresponds to theoretical prediction. (*Source:* Krupenkin, T., S. Yang, and P. Mach. 2003. *Applied Physics Letters*, 82(3), 316–318. With permission.)

in direct contact with the ground electrode beneath the dielectric substrate (Figure 5.26b). Multiple control electrodes positioned beneath or embedded within the dielectric allowed active control over the position and curvature of the droplet. Applying the same biasing voltage to all control electrodes would change the contact angle of the droplet uniformly and thus its curvature.

By applying a biased voltage to control electrodes (such as $V_1 > V_3$; $V_2 = V_4$ in Figure 5.26a), one could initiate the motion of the droplet toward the higher voltage electrode and thus adjust the droplet position. In principle, the liquid droplet can be surrounded by its vapor, but it is often preferable to completely immerse the droplet into a secondary liquid that is immiscible with the liquid that forms the microlens.

Figure 5.27 indicates good agreement between the predicted and measured values of the focal length for this tunable lens for voltages up to ~80 V. At higher voltages, the experimental values of the focal length saturated, while the theoretical curve continued to grow. This discrepancy, however, could be easily explained by noting that at around 80 V the diameter of the microlens contact spot started to approach the outer diameter of the control electrode, thus preventing further spread of the droplet.

To investigate the time response of the microlens, a square wave voltage alternating between 0 and 80 V was applied uniformly to all four electrodes. This caused the microlens to alternate its shape between the initial unperturbed shape at 0 V and a spread altered shape at 80 V. The shape of the microlens was observed to follow the driving voltage as long as the square wave period was above approximately 5 msec.

By utilizing the split electrodes of Krupenkin [18], Yang et al. used electrowetting to precisely align the liquid microlens on the substrate. The liquid

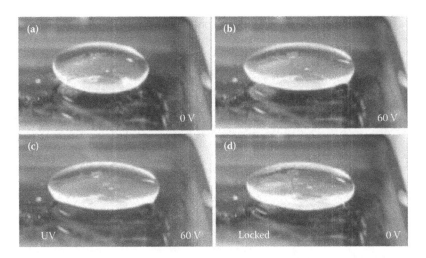

FIGURE 5.28
Photopolymerizable liquid microlens immersed in silicone oil. (a) $V = 0$ V. (b) $V = 60$ V. (c) Exposure to UV light at $V = 60$ V. (d) After UV exposure and removal of applied voltage. (*Source:* Yang, S., T.N. Krupenkin, P. Mach et al. 2003. *Advanced Materials,* 15(11), 940–943. With permission.)

microlens was exposed to UV light to initiate polymerization and thus lock its position and shape. The solidified microlens was then thermally and mechanically robust and able to retain indefinitely its shape, size, and position without applied voltage [20].

A photosensitive polymerizable liquid consisting of NOA 72 optical adhesive and 5.01 wt% of 1-ethyl 3-methyl 1H-imidazolium tetrafluoroborate, was used as a liquid lens. The lens was immersed into DMS-T00 silicone oil to obtain a larger contact angle. DMS-T00 has low surface energy to minimize surface hysteresis and is immiscible with NOA 72 and the salts, thereby ensuring good optical clarity before and after UV exposure.

When 60 V was applied to all four electrodes (Figure 5.26a), the liquid microlens spread, as shown in Figure 5.28a, and a response time of ~100 msec was observed. The lens was then exposed to a broad band UV light while the voltage was still on to fix the lens alignment (Figure 5.28c). The voltage could be removed afterward. Figure 5.28d shows that the lens retained its shape and position with only a small amount of shrinkage even though the liquid was partially cured.

To overcome the evaporation issue, Kuiper et al. demonstrated a liquid lens consisting of two immiscible liquids inside a chamber for miniature cameras [21]. Figures 5.29a and b show a cross section of the variable lens without and with voltage, respectively.

Kuiper's group also described the design and construction of a camera module for use in a mobile phone by using this variable focus lens [21]. Figure 5.30 shows a cross section of the module and photographs of the

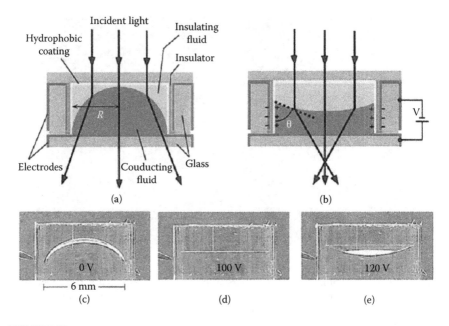

(a)   (b)

(c)   (d)   (e)

FIGURE 5.29
(a) Cross section of electrowetting-based variable lens in cylindrical glass housing. Transparent electrodes were formed of 50 nm ITO. Insulator was a 3 mm parylene-N layer; 10-nm hydrophobic top coating was AF1600 dip-coated fluoropolymer (Dupont Chemicals, Wilmington, DE). Top and bottom glass plates were glued onto glass cylinder with epoxy. (b) When voltage was applied, charges accumulated in wall electrode and opposite charges collected near solid–liquid interface in conducting liquid. Resulting electrostatic force effectively lowered the solid–liquid interfacial tension and contact angle θ. (c) through (e) Video frames of 6 mm diameter lens at 0, 100, and 120 V. (*Source:* Kuiper, S. and B.H.W. Hendriks. 2004. *Applied Physics Letters*, 85(7), 1128–1135. With permission.)

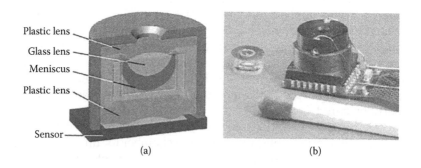

(a)   (b)

FIGURE 5.30
(a) Optical design of camera module containing liquid lens. (b) Assembled camera module and liquid lens. (*Source:* Kuiper, S. and B.H.W. Hendriks. 2004. *Applied Physics Letters*, 85(7), 1128–1135. With permission.)

camera with the electrowetting-based lens. The height of the lens stack was 5.5 mm measured from the image sensor. The f number of the lens was 2.5 with a field view of 60 degrees. A commercially available VGA CMOS sensor (Philips OM6802) of 640 × 480 pixels with a size of 5.0 × 5.0 μm² was used. The entrance pupil diameter was 5.43 mm, and the focal length was adjustable between 2.85 mm for objects at 2 cm and 3.55 mm for objects at infinity.

The electrowetting cell was set between two plastic injection molded lenses. A flat glass plate closed the cylinder on the oil side and a truncated glass sphere mounted on a thin metal diaphragm closed the side of the salt solution. The outer diameter of the cylinder was 4 mm, the inner diameter 3 mm, and the height was 2.2 mm. The flexible metal diaphragm compensated part of the mismatch in thermal expansion between the liquids and the cylinder. The achromatized lens stack had a high optical quality. The camera was able to focus faster than the refresh rate of the CMOS sensor.

Liu et al. reported a double-ring electrode design [22] for electrowetting-based lenses. As illustrated in Figure 5.31, the liquid lens they developed was based on planar electrodes that could be arranged on the same plane without vertical or out-of-plane wall electrodes. This was realized by the unique feature of double-ring shaped electrodes beneath the surface. The outer ring electrode, when applied with an electric potential, could electrowet the area above it and change the surface property. This would provide an initial boundary to confine the insoluble oil droplet. If this voltage was not applied prior to the placement of the oil, it could easily disperse on the hydrophobic surface.

The inner ring, on the other hand, was the actuation electrode of the lens. When an electric voltage was applied to the inner electrode, the surface above it became hydrophilic and attracted the surrounding aqueous solution. The

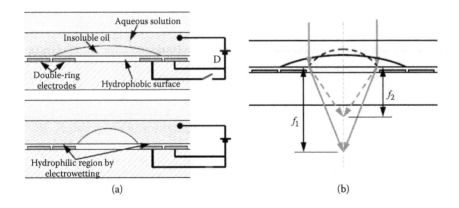

(a)   (b)

FIGURE 5.31
Planar liquid lens design. (a) Curvature change of liquid lens by electrowetting. (b) Focal length change of liquid lens. After application of DC voltage, insoluble oil droplet becomes more curved and focal length shortens from f_1 to f_2. (*Source:* Liu, C.X., J. Park, and J.W. Choi. 2008. *Journal of Micromechanics and Microengineering*, 18(3). With permission.)

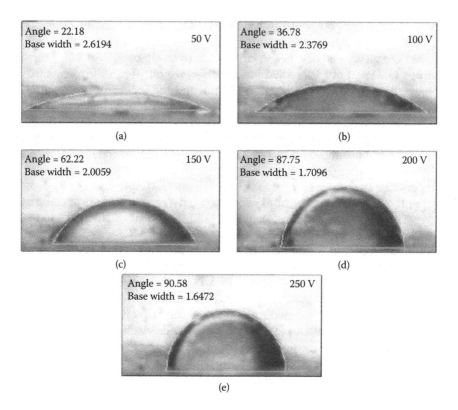

FIGURE 5.32
Images of silicone oil droplet under DC voltages: (a) 50 V, (b) 100 V, (c) 150 V, (d) 200 V, and (e) 250 V. Contact angle changes from about 22 degrees (error: ±1°) at 50 V to ~88 degrees at 200 V. For voltages above 200 V, contact angle shows saturation, changing by fewer than 3 degrees in 50 V of increasing voltage. Base width shown in millimeters. (*Source:* Liu, C.X., J. Park, and J.W. Choi. 2008. *Journal of Micromechanics and Microengineering,* 18(3). With permission.)

aqueous solution deformed the shape of the confined oil droplet, resulting in a change in the focal length of the optical lens.

Figure 5.32 shows the change in contact angle of the droplet at different DC voltages [22]. A wide range of changes was observed. When voltage increased from 50 to 200 V, the contact angle increased by 65 degrees. The change in contact angle slowed dramatically for voltages above 200 V, increasing by fewer than 3 degrees when voltage rose to 250 V. The saturation of the increase in the contact angle reflects a natural thermodynamic limit rather than a deflective property of the structure. This saturation limit is determined by material properties of liquids and substrate [23].

5.5.4 Electrowetting-Based Microlens on a Flexible Curvilinear Surface

The electrowetting-based lenses discussed were produced on flat and mostly rigid substrates. Li et al. extended the technology to flexible polymer surfaces

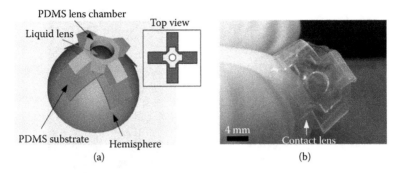

FIGURE 5.33

(a) Electrowetting-based liquid lens fabricated on curved surface. (b) Such a lens made on a commercial contact lens. (*Source:* Li, C. and H. Jiang. 2012. *Applied Physics Letters,* 100(23), 231105. With permission.)

[24]. Figure 5.33a shows the electrowetting-based tunable liquid lens on a flexible substrate. The fabrication of such a lens takes advantage of the transfer technique discussed in Chapter 3. The lens structure was first formed as an island housing the lens on a thin polymer substrate. Both the housing island and the substrate could be made of PDMS. The whole structure could then be wrapped conformally onto a curvilinear surface.

Figure 5.33b shows such a lens made onto a commercial contact lens. The peculiar attribute of flexibility imposes stricter requirements on fabrication processes compared to those applying to flat rigid substrates. For example, a low temperature fabrication process is required to reduce the stress on the PDMS substrate and avoid the damage to PDMS during fabrication. For this reason, the selections of dielectric layer and hydrophobic coating are critical. For instance, SiO_2 was better suited as a dielectric layer than SiN because of its lower stress. Trichlorosilane was better suited than Teflon as a hydrophobic coating since it required lower process temperature. It is worth noting that electrowetting has found many applications beyond liquid microlenses. Hou et al. demonstrated that electrowetting could be used to manipulate optical films [25].

5.6 Electrochemically Activated Adaptive Liquid Microlenses

Lopez et al. devised a surface tension-bounded liquid lens with a pinned contact line that was activated using electrochemistry [26]. The lens was formed by the free surfaces of a liquid overfilling a circular hole of radius R, as shown in Figure 5.34. The contact circles of the liquid were pinned by a non-wetting material; this in turn aligned the optical axis with the center of the orifice. Surface tension was made to change on one capillary surface

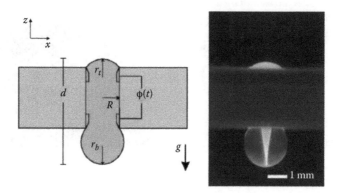

FIGURE 5.34
Liquid lens showing electrode embedded at each contact circle. Photo on right was taken with thin laser light sheet illuminating liquid lens from above. For visualization, fluorescein dye at 4 ppm concentration was dissolved in double-distilled water. Illumination was from a 488 nm line of an argon ion laser. Yellow filter was used on the camera lens. (*Source:* Lopez, C.A., C.C. Lee, and A.H. Hirsa. 2005. *Applied Physics Letters,* 87(13), 134102. With permission.)

relative to the other by utilizing a redox surfactant (FTMA) solution as the lens medium.

The radius of curvature at the tip of each free surface and the maximum thickness of the liquid define the geometric properties of the lens. To calculate the equilibrium configuration of the liquid lens, the authors used the Young-Laplace equation to solve for the top and bottom capillary surfaces, coupled through the hydrostatic pressure of the liquid column with height d and constrained by a particular total volume of liquid. In non-dimensional form, the Young-Laplace equation is

$$\Delta p = 2H - Bz \tag{5.10}$$

where the pressure difference across the capillary surface is scaled by σ_0/R and σ_0 is the surface tension of pure water. H is the mean curvature of a given surface scaled by R. The Bond number B is $\rho g R^2/\sigma_0$, ρ is the liquid density, g is the gravitational acceleration, and z is the axial coordinate scaled by R. For a thick liquid lens, the focal length f is given by

$$\frac{1}{f} = \left(\frac{n'}{n}-1\right)\left(\frac{1}{r_b} + \frac{1}{r_t + \left(\frac{n'}{n}-1\right)d}\right) \tag{5.11}$$

where n and n' are the refractive indices of air and water, respectively, r_t and r_b are the radii of curvature of the top and bottom capillary surfaces

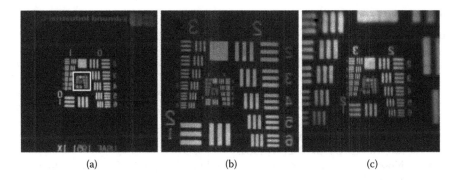

FIGURE 5.35
(a) and (b) Images of standard target obtained using liquid lenses with pure water. (c) Images taken with FTMA solution after four cycles of activation. White box in (a) delineates region imaged in (b). Volumes and focal lengths were $V_{total} = 5.37$ V_{sphere}, $f/R = 5.02$ for (a), $V_{total} = 5.18$ V_{sphere}, $f/R = 2.85$ for (b), and $V_{total} = 5.21$ V_{sphere}, $f/R = 2.39$ for (c). (*Source:* Lopez, C.A., C.C. Lee, and A.H. Hirsa. 2005. *Applied Physics Letters*, 87(13), 134102. With permission.)

evaluated at their respective intersection with the optical axis, and d is the thickness of the lens, as shown in Figure 5.34.

Activation of the liquid lens was achieved by using a water soluble ferrocenyl surfactant whose surface activity could be controlled electrochemically [27]. Application of a voltage difference across the lens produced a reversible reduction–oxidation process that modified the surface activity of the surfactant. The surface tension increased where oxidation occurred and decreased where reduction occurred with a magnitude of ~8 dyn/cm. The total change in surface tension (16 dyn/cm) was more than one-fifth of the surface tension of pure water. This difference in surface tension modified the response of the system, resulting in a redistribution of the liquid.

The orientation of the anode and the cathode (top versus bottom electrode) determined large or small focal length changes. Where voltage was applied and the corresponding changes in surface tension had the net effect of equalizing V_{top} and V_{bottom}, large volume displacement occurred but focal length changed very little since r_t, r_b, and d did not change significantly. On the other hand, if the applied voltage and resultant changes in surface tensions drove either V_{top} or V_{bottom} toward 0, the focal length increased by a factor of ~3 for the air–water system.

Lopez et al. also tested the resolution of the liquid lens by imaging a standard target. A U.S. Air Force 1951 glass slide was imaged through the liquid lens and photographed. Figure 5.35b shows the smallest features seen with a pure water liquid lens of 45.3 line pairs per millimeter in both horizontal and vertical directions. The maximum resolution achieved with an FTMA lens (Figure 5.35c) was 22.62 line pairs per millimeter.

References

1. S. Sato, "Liquid-Crystal Lens-Cells with Variable Focal Length," *Japanese Journal of Applied Physics*, vol. 18, pp. 1679–1684, 1979.
2. S. T. Kowel, D. S. Cleverly, and P. G. Kornreich, "Focusing by Electrical Modulation of Refraction in a Liquid-Crystal Cell," *Applied Optics*, vol. 23, pp. 278–289, 1984.
3. S. T. Kowel, P. Kornreich, and A. Nouhi, "Adaptive Spherical Lens," *Applied Optics*, vol. 23, pp. 2774–2777, 1984.
4. T. Nose, S. Masuda, and S. Sato, "A Liquid-Crystal Microlens with Hole-Patterned Electrodes on Both Substrates," *Japanese Journal of Applied Physics Part 1-Regular Papers Short Notes & Review Papers*, vol. 31, pp. 1643–1646, May 1992.
5. Y. Choi, J. H. Park, J. H. Kim, and S. D. Lee, "Fabrication of a focal length variable microlens array based on a nematic liquid crystal," *Optical Materials*, vol. 21, pp. 643–646, Jan 2003.
6. N. A. Riza and M. C. Dejule, "Three-terminal adaptive nematic liquid-crystal lens device," *Optics Letters*, vol. 19, pp. 1013–1015, 1994.
7. C. C. Cheng, C. A. Chang, and J. A. Yeh, "Variable focus dielectric liquid droplet lens," *Optics Express*, vol. 14, pp. 4101–4106, May 2006.
8. H. W. Ren, Y. H. Fan, and S. T. Wu, "Liquid-crystal microlens arrays using patterned polymer networks," *Optics Letters*, vol. 29, pp. 1608–1610, Jul 2004.
9. H. W. Ren, Y. H. Fan, and S. T. Wu, "Tunable Fresnel lens using nanoscale polymer-dispersed liquid crystals," *Applied Physics Letters*, vol. 83, pp. 1515–1517, Aug 2003.
10. G. Q. Li, P. Valley, M. S. Giridhar, D. L. Mathine, G. Meredith, J. N. Haddock, B. Kippelen, and N. Peyghambarian, "Large-aperture switchable thin diffractive lens with interleaved electrode patterns," *Applied Physics Letters*, vol. 89, p. 141120, Oct 2006.
11. N. Binh-Khiem, K. Matsumoto, and I. Shimoyama, "Polymer thin film deposited on liquid for varifocal encapsulated liquid lenses," *Applied Physics Letters*, vol. 93, p. 124101, Sep 2008.
12. C. C. Cheng and J. A. Yeh, "Dielectrically actuated liquid lens," *Optics Express*, vol. 15, pp. 7140–7145, Jun 2007.
13. H. Ren and S. T. Wu, "Tunable-focus liquid microlens array using dielectrophoretic effect," *Optics Express*, vol. 16, pp. 2646–2652, Feb 2008.
14. L. G, "Relations entre les ph´enom`enes ´electriques et capillaires," *Ann. Chim. Phys.*, vol. 5, p. 494, 1875
15. F. Mugele and J.-C. Baret, "Electrowetting: from basics to applications," *Journal of Physics: Condensed Matter*, vol. 17, p. R705, 2005.
16. C. B. Gorman, H. A. Biebuyck, and G. M. Whitesides, "Control of the Shape of Liquid Lenses on a Modified Gold Surface Using an Applied Electrical Potential across a Self-Assembled Monolayer," *Langmuir*, vol. 11, pp. 2242–2246, Jun 1995.
17. B. Berge and J. Peseux, "Variable focal lens controlled by an external voltage: An application of electrowetting," *The European Physical Journal E*, vol. 3, pp. 159 - 163, 2000.
18. T. Krupenkin, S. Yang, and P. Mach, "Tunable liquid microlens," *Applied Physics Letters*, vol. 82, pp. 316–318, Jan 2003.

19. P. G. de Gennes, "Wetting: statics and dynamics," *Reviews of Modern Physics*, vol. 57, pp. 827–863 1985.
20. S. Yang, T. N. Krupenkin, P. Mach, and E. A. Chandross, "Tunable and latchable liquid microlens with photopolymerizable components," *Advanced Materials*, vol. 15, pp. 940–943, Jun 2003.
21. S. Kuiper and B. H. W. Hendriks, "Variable-focus liquid lens for miniature cameras," *Applied Physics Letters*, vol. 85, pp. 1128–1130, Aug 2004.
22. C. X. Liu, J. Park, and J. W. Choi, "A planar lens based on the electrowetting of two immiscible liquids," *Journal of Micromechanics and Microengineering*, vol. 18, p. 035023, Mar 2008.
23. A. Quinn, R. Sedev, and J. Ralston, "Contact Angle Saturation in Electrowetting," *The Journal of Physical Chemistry B*, vol. 109, pp. 6268–6275, 2005.
24. C. Li and H. Jiang, "Electrowetting-driven variable-focus microlens on flexible surfaces," *Applied Physics Letters*, vol. 100, p. 231105, 2012.
25. L. Hou, N. R. Smith, and J. Heikenfeld, "Electrowetting manipulation of any optical film," *Applied Physics Letters*, vol. 90, p. 251114, Jun 2007.
26. C. A. Lopez, C. C. Lee, and A. H. Hirsa, "Electrochemically activated adaptive liquid lens," *Applied Physics Letters*, vol. 87, p. 134102, Sep 2005.
27. B. S. Gallardo, V. K. Gupta, F. D. Eagerton, L. I. Jong, V. S. Craig, R. R. Shah, and N. L. Abbott, "ELectrochemical principles for active control of liquids on sub-millimeter scales," *Science*, vol. 283, pp. 57–60, Jan 1999.

6

Mechanically Driven Tunable Microlenses

In the last chapter, we discussed a few methods to form tunable focus liquid lenses. The underlying tuning mechanism was usually electrical actuation. Another category of actuation for focal length tuning of liquid lenses is mechanical force. This chapter will describe microlenses tuned by a variety of mechanical methods. These mechanically tunable devices include lenses consisting of a thin polymer diaphragm in a chamber or a microchannel, actuated by varying the pneumatic pressure in the chamber or channel; lenses formed from liquid–liquid interfaces, actuated by environmentally stimuli-responsive hydrogels; lenses actuated via deformation of their apertures; lenses formed from swellable hydrogels; lenses tuned by changing their apertures; lenses tuned by snapping surfaces; oscillating lens arrays driven by sound waves; and strain-responsive microlens arrays.

6.1 Introduction

Many mechanically tunable microlenses are designed as circular chambers covered by thin flexible membranes. The membrane deforms when pressure is applied to the lens liquid through external actuation. In addition, some microlenses are formed through air–liquid or liquid–liquid interfaces (immiscible, with different refractive indices). These interfaces may also be varied by applying pressure that adjusts the radii of curvatures of the spherical membranes in air–liquid devices and the focal lengths of the microlenses formed via liquid–liquid interfaces.

6.2 Thin Glass Membranes

The initial design of a tunable microlens in a chamber covered with a thin membrane was inspired by the crystalline lenses of human eyes. Ahn et al. reported a variable focusing lens consisting of a glass diaphragm, a silicon substrate with microchannels for fluid manipulation, and a bottom glass substrate [1]. Glass was placed on the silicon substrate by anodic bonding.

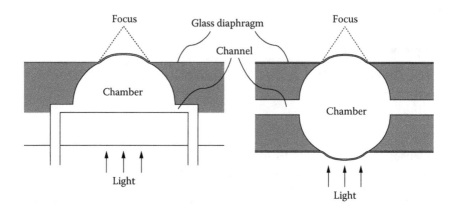

FIGURE 6.1
Variable focusing lenses with glass diaphragms. (*Source:* Ahn, S.H. and Y.K. Kim. 1999. *Sensors and Actuators A*, 78(1), 48–53. With permission.)

The chamber and microchannels for fluid manipulation were formed on the silicon side by photolithography.

The lens consisted of a silicon structure with bonded glass diaphragm (Figure 6.1a). Two of the same lens parts were later bonded face to face to complete a lens with double glass diaphragms (Figure 6.1b). The chamber between the two glass diaphragms was filled with a working fluid. Oils for refractive index matching (e.g., silicone oil) could be used as working fluids. The fluid was pumped into or out of the chamber through the microchannel via an external micropump; during this process the oil pressure varied, changed the curvature of the lens, and caused a shift in the focal plane. In their design [1], the variable focusing lens had a 1 cm × 1 cm square diaphragm. The thickness of the glass diaphragm was 40 μm.

Due to the relatively high Young's modulus of the glass diaphragm, large pressure was required to deflect the membrane. As shown in Figure 6.2, when the deflection does occur, it is quite sensitive. A small deformation may lead to a relatively large change in focal length. Hence, the pressure must be well controlled for the desired operation range of the lens.

6.3 Polymer Membranes

6.3.1 Single Lenses

To achieve larger deformation of the membranes and the resulting curved surfaces, soft materials such as polymers have replaced the hard glass. Zhang et al. first reported a fluidic adaptive lens consisting of a polydimethylsiloxane (PDMS) fluidic chamber covered by a thin (60 μm) PDMS membrane and bonded to a thin (150 μm) handling glass slide [2].

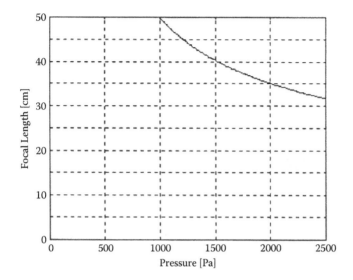

FIGURE 6.2
Plot of focal length of glass diaphragm-based variable focus liquid lens versus driving pressure. (*Source:* Ahn, S.H. and Y.K. Kim. 1999. *Sensors and Actuators A*, 78(1), 48–53. With permission.)

Werber et al. [3] presented a microfluidic microlens employing a 50 μm thick pressure-actuated PDMS membrane structured on a silicon fluidic chip. The lenses were 400 μm in diameter and could be both planoconvex and planoconcave by varying the pressure of the liquid through the fluidic system and changing the shape of the membrane and thus altering the optical power of the lens as shown in Figure 6.3.

Chronis et al. reported a similar lens array with a thinner PDMS membrane of 40 μm and smaller lens diameter of 200 μm [4]. They also used finite element analysis software (ANSYS) to simulate the mechanical deformation of the membrane under uniform pressure. Figure 6.4a shows a cross sectional view of a single deformed membrane simulated by ANSYS. An important finding was the difference between the maximum deflection of the outer (top) surface of the membrane and the inner (bottom) one. The inner surface of the membrane deflects more than the outer one, resulting in a smaller radius of curvature. Figure 6.4b illustrates the outer and inner radius of curvature measured and calculated from the interferometric lens profile and ANSYS simulations, respectively [4].

Besides the planar surfaces, predefined convex surfaces can be utilized to provide high numerical apertures [5,6]. Jeong et al. used a molded concave PDMS surface for the lens [5], while Chen et al. used a concave PDMS surface as a mold to make a convex lens [6]. Figure 6.5 shows a microdoublet lens consisting of a tunable liquid-filled microlens and a solid negative elastomer lens of different refractive indices in combination [5]. Each liquid-filled microlens with a variable and a fixed lens curvature was connected by microfluidic networks.

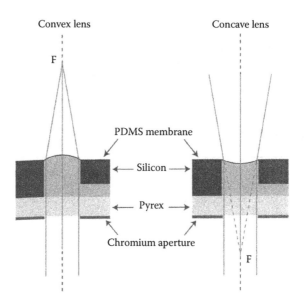

FIGURE 6.3

Cross section of PDMS membrane-based microfluidics microlens showing implementation as planoconvex and planoconcave lenses. (*Source:* Werber, A. and H. Zappe. 2005. *Applied Optics,* 44(16), 3238–3245. With permission.)

The curvature of the fixed lens was defined by the mold for the fabrication of the lenslet. The curvature of the variable lens was dynamically tuned with the deflection of a thin polymer membrane under pneumatic control via microfluidic channels.

Feng et al. introduced a meniscus–biconvex PDMS lens [7]. As shown in Figure 6.6, the chamber was controlled by a fluidic pressure and the walls of the lens chamber formed a complex PDMS–liquid–PDMS lens system [7]. The meniscus lens part could reduce the overall focal length without introducing extra spherical aberration or even reduce light beam distortion. Meanwhile, the biconvex lens part could make a collimated beam of light converge after passing the lens and obtain image magnification. Moreover, the focus of the flexible meniscus and biconvex lens with non-spherical surfaces could be tuned again by exerting fluid pressure. The authors also showed the relationship of the exerted fluid pressures and focal distances and field of view (FOV) values (Figure 6.7). The results revealed a quadratic relation between the pressure values and the focal distances.

By utilizing two free membranes, both Agarwal et al. [8] and Moran et al. [9] reported biconvex variable focal length microlens systems with large FOV and numerical aperture values. The system is illustrated in Figure 6.8.

Yu et al. presented an intriguing design based on the standard liquid-filled microlens with bifocal capability: one lens is formed by the inner portion and one is created by the peripheral region [10]. The underlying mechanism

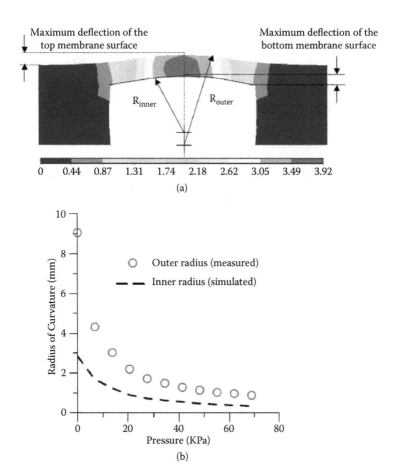

(a)

(b)

FIGURE 6.4
(a) Cross-sectional view of simulated PDMS membrane pressurized at 20 KPa. Finite element simulation reveals two radii of curvature. (b) Outer and inner radii of curvature versus applied pressure. (*Source:* Chronis, N., G.L. Liu, K.H. Jeong et al. 2003. *Optics Express*, 11(19), 2370–2378. With permission.)

of this microlens is the same as in the previous designs. A suspended PDMS membrane and substrate form a closed cavity defining the microlens aperture. The cavity is connected to an external pumping system via an integrated microfluidic microchannel through which fluid could be introduced into the cavity.

By changing the fluid volume and cavity pressure, the membrane was forced to deflect. This created a surface contour of the lens. In contrast to previous designs, the authors adopted a membrane structure with different thicknesses for the central and peripheral regions, as shown in Figure 6.9. The resultant deformations and radii of curvature at those two regions under the same pressure were thus different. As a result, they generated two separate

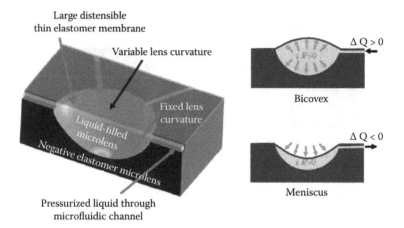

FIGURE 6.5
Basic configurations of tunable microdoublet lens consisting of tunable liquid-filled microlens and solid negative lens of different refractive indices ($n_{elastomer}$ and n_{liquid}, respectively) acting in combination. (*Source:* Jeong, K.H., G.L. Liu, N. Chronis et al. 2004. *Optics Express*, 12(11), 2494–2500. With permission.)

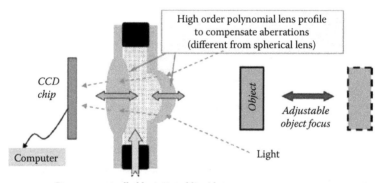

FIGURE 6.6
Conceptual diagram of flexible meniscus–biconvex lens system with focus tuning functions. (*Source:* Feng, G.H. and Y.C. Chou. 2009. *Sensors and Actuators A*, 156(2), 342–349. With permission.)

lenses with different focal lengths. The structural parameters for this bifocal lens are listed in Table 6.1.

The authors utilized the ANSYS finite element analysis software to simulate the relationship of membrane deformation, focal length, and applied pressure [10]. The simulation results for the cross-sectional contour of the deformed membrane under pressures ranging from 200 Pa to 9.5 kPa are shown in Figure 6.10a. As comparison, the profile of a microlens with a uniform thickness of 90 μm under 5 kPa appears in Figure 6.10b. Focal lengths were calculated by inputting the surface profiles of the lenses into Zemax software.

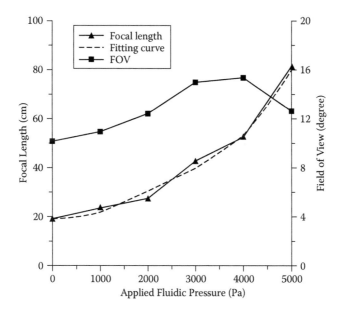

FIGURE 6.7
Relationship of fluidic pressure, focal distances, and FOV values for flexible meniscus–biconvex lens. (*Source:* Feng, G.H. and Y.C. Chou. 2009. *Sensors and Actuators A*, 156(2), 342–349. With permission.)

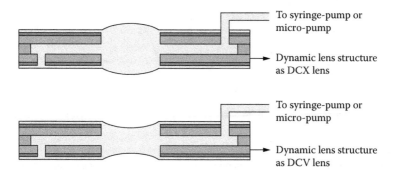

FIGURE 6.8
Bi-membrane lens system. (*Source:* Agarwal, M., R.A. Gunasekaran, P. Coane et al. 2004. *Journal of Micromechanics and Microengineering*, 14(12), 1665–1673. With permission.)

The focal length at the central region could be adjusted gradually from ~46.05 mm to nearly 10.43 mm when the pressure was changed from 200 Pa to 9.5 kPa. The tunable range was 57.34 to 15.08 mm in the peripheral region. Figure 6.11 shows the function of the bifocal lens [10]. A collimated beam is transmitted through the lens. Two focused bright spots due to the bifocal length of the lens were clearly observed.

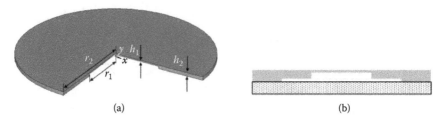

(a)   (b)

FIGURE 6.9
(a) Schematic and (b) side view of membrane structure for bifocal microlens design. (*Source:* Yu, H.B., G.Y. Zhou, F.K. Chau et al. 2009. *Optics Express*, 17(6), 4782–4790. With permission.)

TABLE 6.1

Structure Parameters of Membrane for Bifocal Microlens

Part	r_1	r_2	h_1	h_2
Parameter	1.25 mm	2.5 mm	30 μm	90 μm

Source: Yu, H.B., G.Y. Zhou, F.K. Chau et al. 2009. *Optics Express*, 17(6), 4782–4790. With permission.

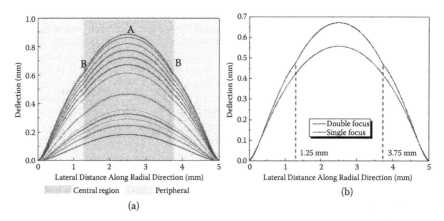

FIGURE 6.10
Simulation results are shown for the membrane deformation in the bifocal lens. (a) Membrane deformation under different pressures (From bottom to top, the pressure value are 200 Pa, 400 Pa, 600 Pa, 800 Pa, 1 kPa, 2 kPa, 3 kPa, 4 kPa, 5 kPa, 6 kPa, 7 kPa, 8 kPa, 9 kPa and 9.5 kPa, respectively); (b) The comparison of membrane deformation under 5 kPa pressure in a single-focus and double-focus design, respectively. Reprinted from [10] with permission from Optical Society of America.

Xiong et al. proposed a microlens possessing both tunable focal length and variable light transmission [11]. The structure of the device is depicted in Figure 6.12. It contains a liquid cell sealed with a 100 μm thick membrane layer of elastic PDMS on its upper surface. The cell is connected to a syringe via a tube for tuning. The liquid filled into the liquid cell in this work was an aqueous solution of 2.5% poly-N-isopropylacrylamide (PNIPAAm).

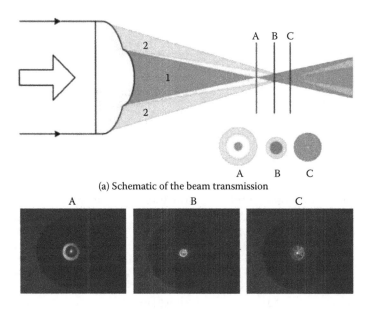

(a) Schematic of the beam transmission

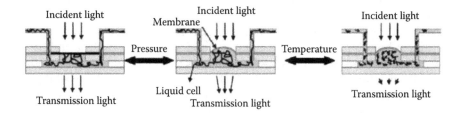

FIGURE 6.11
Beam transmission through bifocal microlens. (a) Beam transmission. (b) Image of beam captured at different positions along the transmission path. (*Source:* Yu, H.B., G.Y. Zhou, F.K. Chau et al. 2009. *Optics Express*, 17(6), 4782–4790. With permission.)

FIGURE 6.12
Microlens device with focal length tunability and light transmission control. (*Source:* Xiong, G.R., Y.H. Han, C. Sun et al. 2008. Liquid microlens with tunable focal length and light transmission. *Applied Physics Letters*, 92(24). With permission.)

By controlling the pressure in the chamber, the liquid device acted as a convergent microlens. The radius of curvature of the PDMS membrane was controlled by the volume of liquid injected into the cell. The focal length varied from 1.98 mm to positive infinity. In addition to the tuning in the focal length, light transmission of the microlens was also controllable. This was because PNIPAAm is a thermosensitive polymer whose polymeric chains exhibit changes in conformation around its lower critical solution temperature (LCST). At temperatures below the LCST, the intermolecular hydrogen bonding between the amide group of the PNIPAAm chain and the water molecules was predominant.

TABLE 6.2

Comparison of Performances of Pneumatically Driven Polymer Microlenses

Author and Reference	Focal Length (mm)	Pneumatic Pressure or Volumetric Change	f Number	Field of View (degrees)
Zhang et al. [2]	8 to 20	0 to 15 kPa (front lens); 0 to 30 kPa (back lens)	0.06 to 0.24	10 to 80
Werber and Zappe [3]	1 to 18	4 to 54 kPa	n/a	n/a
Chronis et al. [4]	0.5 to 2 (oil)	0 to 35 kPa	n/a	n/a
Jeong et al. [5] Norland [63]	1.8 to 6.2			
Chen et al. [6]	3.815 to 10.64	0 to 70 μL	0.09 to 0.24	n/a
Feng et al. [7]	191 to 808	0 to 5 kPa	n/a	10 to 15
Agarwal et al. [8]	75.9 to 3.1 −75.9 to −3.3	0 to 14.5 μL 0 to 13.5 μL	0.61 to 5.02	0.12 to 61 (double convex); 7 to 69 (double concave)
Moran et al. [9]	2.9 to 7.5	61 to 146.9 Pa	0.13 to 0.35	n/a
Yu et al. [10]	10 to 58	0.2 to 9.5 kPa	Varied	n/a
Xiong et al. [11]	1.9 to 3.5	11 to 21 μL	n/a	n/a

Note: n/a = Not available.

The polymer chains then expanded in water, resulting in a transparent liquid. When the temperature exceeded the LCST, the PNIPAAm chains collapsed and formed clusters of polymer chain coils due to the intramolecular hydrogen bonding between the isopropyl and amide groups. The coiling of the polymer chains strongly scattered the incident light, reducing the transmittance through the liquid lens. The value of transmittance was maintained at 94% below 34.5°C and dramatically decreased to 10% above this temperature. All changes were reversible [11]. Note that the focal length and the transmittance were controlled by the volume of liquid and the temperature separately and independently.

To compare the existing pneumatically driven tunable microlens discussed above, we summarized some important parameters in Table 6.2.

6.3.2 Microlenses in Groups

The microlenses discussed to this point utilize one tuning chamber or cell. By adjusting the fluidic pressure inside the lens chamber or cell, one can turn a fluidic lens into various other types such as planoconvex, planoconcave, biconvex, biconcave, positive meniscus, and negative meniscus [8,9]. For each lens type, the optical parameters including f and *NA* can be freely adjusted.

Zhang et al. formed a fluidic adaptive lens consisting of two back-to-back PDMS fluidic chambers, each covered by a thin (60 μm) PDMS membrane [12]. A thin glass slide (150 μm) was sandwiched between the two lens

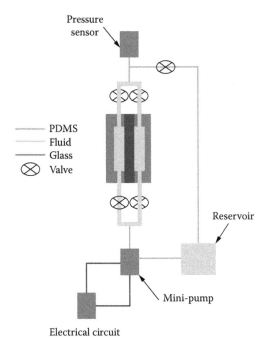

Pressure sensor

PDMS
Fluid
Glass
⊗ Valve

Reservoir

Mini-pump

Electrical circuit

FIGURE 6.13
Fluidic control system for fluidic adaptive lenses. (*Source:* Zhang, D.Y., N. Justis, and Y.H. Lo. 2004. Fluidic adaptive lens of transformable lens type. *Applied Physics Letters*, 84(21), 4194–4196. With permission.)

chambers and the entire device was assembled via oxygen plasma-activated PDMS–glass bonding. The lens aperture was 20 mm. DI water and 63% sodium chromate were used as the media for the fluidic adaptive lenses. Figure 6.13 is a diagram of the fluidic system used to control the lens properties. By separately controlling the fluidic pressure in the two lens chambers, different types including planoconvex, planoconcave, biconvex, biconcave, positive meniscus, and negative meniscus lenses were achieved with wide ranges of focal length tuning for each type.

Pang et al. proposed a liquid-filled lens design operating as an assembly of two orthogonally and independently controlled adaptive cylindrical lenses [13]. As shown in Figure 6.14, the device consisted of four separate layers of PDMS, and had four layers of liquid-filled chambers that were bonded together. The main functional element in each layer was a rectangular chamber ~250 μm in depth and with a length-to-width ratio of about 5:1.

The chambers in layers 2 and 3 (chambers 2 and 3) were connected and created a single cross-shaped chamber in the middle plane of the device. The chambers in layers 1 and 4 (chambers 1 and 4) were separated from chambers 2 and 3 by two flexible membranes about 200 μm thick. The two

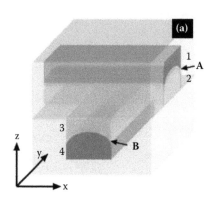
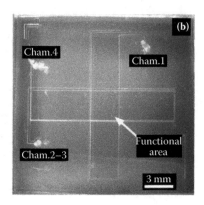

FIGURE 6.14

(a) Assembly of two orthogonal adaptive cylindrical lenses, demonstrating its operation as a positive cylindrical lens along the *y* axis and a negative cylindrical lens along the *x* axis. (b) Micrograph of fabricated device. Three inlets were connected to chambers 1, 2–3 and 4. (*Source:* Pang, L., U. Levy, K. Campbell. 2005. *Optics Express*, 13(22), 9003–9013. With permission.)

membranes were centered with respect to the optical axis of the device parallel to the *z* axis in Figure 6.14.

The cross-shaped chamber in the middle plane was filled with a liquid having a high refractive index n_2. Chambers 1 and 4 were filled with another liquid with a lower refractive index n_1. The optically functional area of the device was in the center where the two membranes overlapped. Differences in the pressures of the liquid in chamber 1 (P_1), and in chambers 2–3 (P_2) caused bending of membrane A.

The bending created a cylindrical interface between the liquids in chambers 1 and 2 that acted as a cylindrical lens modifying the wavefront of light along *y* axis, with a tunable focal length of f_x. The lens was converging if $P_2 > P_1$ and diverging if $P_2 < P_1$. Membrane B operated similarly. The dependence of the focal lengths f_x and f_y on their respective control pressures ΔP_x and ΔP_y, are shown in Figure 6.15 [13]. The dependencies of f_x on ΔP_x and f_y on ΔP_y were very similar, indicating that the two cylindrical lenses were nearly identical.

6.3.3 Variable Focus Liquid Lens by Changing Aperture Size

Ren et al. presented a design to vary the focal length of a liquid lens by changing its aperture size [14,15]. A lever actuator was used to control the movement of rotatable impellers that in turn imparted a pressure on the fluid-filled lens. Thus, the redistributed liquid changed the lens curvature and tuned the focal length.

Figure 6.16a depicts such a liquid lens consisting of a glass plate, a circular periphery seal, and a clear elastic membrane [14]. The circular peripheral seal was wrapped with an elastic membrane to confine the liquid. The key element of the liquid lens was the circular peripheral seal resembling a

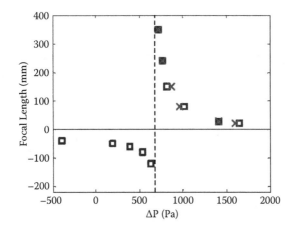

FIGURE 6.15
Focal lengths of the cylindrical lenses focusing light along the x-axis (squares) and y-axis (crosses) are shown as functions of the pressure differences, ΔP_x and ΔP_y, respectively. Reprinted from [13] with permission from *Optical Society of America*.

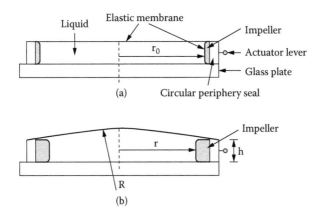

FIGURE 6.16
Cross-sectional view of the liquid lens cell with (a) no focusing effect with aperture radius r_0 and (b) focusing effect with aperture radius $r < r_0$. (*Source:* Ren, H. and S.T. Wu. 2005. *Applied Physics Letters*, 86(21), 211107. With permission.)

conventional iris diaphragm, as shown in Figure 6.17. The rotatable impellers were also incorporated with the circular peripheral seal. A lever actuator was used to control the movement of the rotatable impellers.

Figure 6.17 (left) shows the situation when the radius of the aperture was r_0, corresponding to when the elastic membrane is flat as shown in Figure 6.16a. When the radius of the aperture was reduced to r_1 ($r_1 < r_0$) by rotating the lever actuator as shown in Figure 6.17b, the liquid was pressed to push the elastic membrane outward. Thus, a positive (or converging) lens was formed

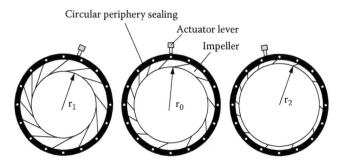

FIGURE 6.17
Top view of circular peripheral seal used in tunable liquid lens described by Ren and Wu. The aperture radii are (left) r_0, (middle) r_1 ($r_1 < r_0$), and (right) r_2 ($r_2 < r_0$). (*Source:* Ren, H. and S.T. Wu. 2005. *Applied Physics Letters*, 86(21), 211107. With permission.)

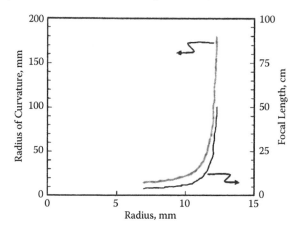

FIGURE 6.18
Simulated radius of curvature R and focal length f as function of lens aperture size r. Thickness of rotatable impellers = 5.5 mm; h = 5.5 mm; r_0 = 12.5 mm; and refractive index of filling liquid n = 1.36. (*Source:* Ren, H. and S.T. Wu. 2005. *Applied Physics Letters*, 86(21), 211107. With permission.)

(Figure 6.16b). If the radius of the aperture was enlarged to r_2 ($r_2 > r_0$) as shown in Figure 6.17c, a negative (or diverging) lens was formed.

Figure 6.18 plots the simulated focal length (right side ordinate) of the resultant tunable lens and the radius of the lens aperture (left side ordinate) [14]. The focal length and the radius of curvature showed a similar trend.

6.3.4 Liquid Lens Based on Pressure-Induced Liquid Redistribution

In the previous section, we introduced a liquid lens whose focal length was varied by pressurizing a circular membrane [14,15]. The reservoir chamber was around the periphery that was wrapped with an elastic membrane. However, the elastic membrane occupied a large area and is not conducive

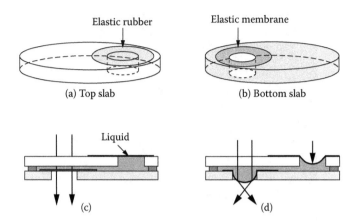

FIGURE 6.19
Structure of liquid lens cell. (a) Top slab. (b) Bottom slab and side view of lens cell in (c) non-focusing and (d) focusing states. (*Source:* Ren, H.W. and S.T. Wu. 2007. *Optics Express*, 15(10), 5931–5936. With permission.)

to lens miniaturization. Ren et al. demonstrated a liquid lens based on pressure-induced liquid redistribution [16].

Figure 6.19 depicts the design and fabrication of the lens reported by Ren and Wu [16]. The lens cell had two apertures and was flat in the initial non-focusing state. One aperture was sealed with an elastic membrane on the outer side of the top substrate surface and the other aperture was sealed with an elastic membrane on the inner surface of the bottom substrate. These two apertures did not overlap spatially.

Two clear glass or plastic slabs were used as the lens frames. Each slab was drilled with a hole and each hole was sealed with an elastic membrane (Figure 6.19a and b). The two slabs were sandwiched together to form a flat cell. The periphery of the cell was sealed with epoxy glue except for a hole connecting to the chamber. A liquid was injected into the chamber through the hole and then the hole was sealed with glue as well to finish the package.

Figure 6.19c shows a cross-sectional view of the lens cell in a flat state. The liquid was incompressible. When an external pressure was applied to deform the outer elastic rubber inward, the liquid in the lens chamber was redistributed, forcing the inner elastic membrane to swell outward. As shown in Figure 6.19d, the result was a planoconvex lens and the incident light could be focused. To activate the liquid lens, an electrically controlled actuator or mechanical lever could be employed.

6.3.5 Snapping Surfaces for Tunable Microlenses

To accomplish fast tuning time of microlenses, Holmes et al. utilized snap transition to change the status of microlenses [17]. This snap transition was based on the onset of an elastic snap-through instability similar to

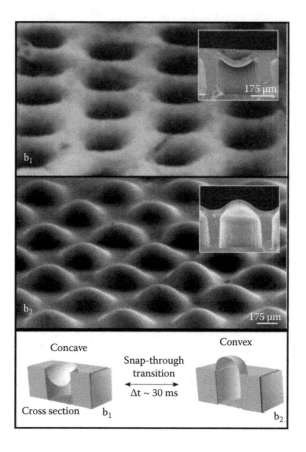

FIGURE 6.20

Tunable microlenses formed by snap-through transition. Top: Concave microlenses. Middle: Convex microlenses. Bottom: Snapping process from concave to convex. (*Source:* Holmes, D.P. and A.J. Crosby. 2007. *Advanced Materials,* 19(21), 3589–3593. With permission.)

the capture mechanism of a Venus flytrap plant [18]. The mechanical response rates can be milliseconds, comparable to rates for electrowetting (Section 5.5).

Figure 6.20 shows the snapping microlenses reported by Holmes [17]. If a trigger mechanism was used to develop critical stress in all shells simultaneously, the entire surfaces of shells would change curvature. One way to achieve this switching of multiple lenses from convex to concave was by exposing the responsive surface to oxygen plasma treatment.

This process caused a volumetric decrease on the surface of the polymeric film, triggering the shells to snap from convex to concave to minimize the development of tensile stresses in the outer surface layer. To snap the shells from concave to convex, a triggering mechanism that caused volumetric expansion could be used. The authors demonstrated one such approach by swelling the elastic polymer network with an organic solvent to develop an

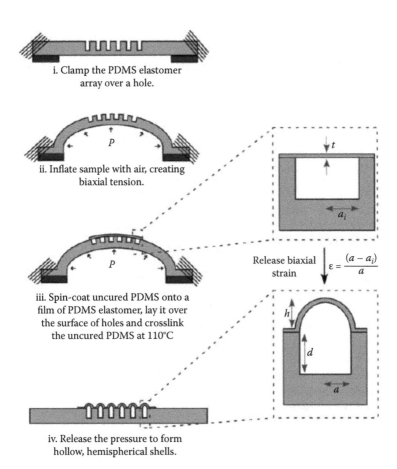

i. Clamp the PDMS elastomer
array over a hole.

ii. Inflate sample with air, creating
biaxial tension.

iii. Spin-coat uncured PDMS onto a
film of PDMS elastomer, lay it over
the surface of holes and crosslink
the uncured PDMS at 110°C

iv. Release the pressure to form
hollow, hemispherical shells.

Release biaxial strain $\quad \varepsilon = \dfrac{(a - a_i)}{a}$

FIGURE 6.21
Fabrication of the microlenses reported by Holmes. (*Source:* Holmes, D.P. and A.J. Crosby. 2007. *Advanced Materials*, 19(21), 3589–3593. With permission.)

osmotic stress, similar to the mechanism of a Venus flytrap [17]. Because of the lateral confinement at the edges, the swelling surface would buckle under the compressive stress, and the lenses would snap.

Figure 6.21 shows the fabrication of the active surface structures reported by Holmes [17] using the Euler buckling of plates to generate a controlled array of microlens shells under equibiaxial compressive stress. First, cylindrical posts of photoresist (PR) were photopatterned onto a silicon wafer, followed by micromolding a PDMS elastomer onto the silicon substrate, creating an array of holes. This elastic PDMS elastomer film with an array of holes then underwent equibiaxial strain through an inflation procedure.

A thin film of cross linked PDMS (15 to 60 μm thick) coated with a thin (~1 μm) layer of uncured PDMS was placed on the surface of the strained holes. The assembly was heated to cross link the uncured PDMS and bond the film

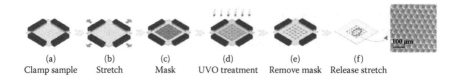

(a)   (b)   (c)   (d)   (e)   (f)
Clamp sample   Stretch   Mask   UVO treatment Remove mask   Release stretch

FIGURE 6.22
(a) through (e) Fabrication of concave microlens array in PDMS film. (f) Scanning electron microscopy image of fabricated concave microlens array. (*Source:* Chandra, D., S. Yang, and P.C. Lin. 2007. *Applied Physics Letters*, 91(25). With permission.)

to the array of holes while under equibiaxial tension. Releasing the tension caused an equibiaxial compressive strain to develop in the thin PDMS coating. The strain in turn caused the circular plate of PDMS on the surface of each hole to buckle, thus creating an array of convex microlenses.

The microlenses of Holmes [17] were bistate types. Chandra et al. utilized a similar mechanism to fabricate a single component, strain-responsive microlens array capable of continuous focus tuning [19]. Figure 6.22 shows the fabrication of the array. A flat PDMS sheet 0.5 mm thick was prepared first. The sheet was clamped at four edges (Figure 6.22a) and then stretched to 20% strain in both planar directions simultaneously (Figure 6.22b).

One side of the stretched PDMS surface was masked with a transmission electron microscopy (TEM) copper grid (Figure 6.22c) with a hexagonally packed hole array (diameter = 37 μm; hole-to-hole gap = 62 μm) for ultraviolet ozone (UVO) treatment for 30 min (Figure 6.22d) to generate a thin silicate layer on the exposed regions.

The area surrounding the TEM grid and the back side of the PDMS film were covered. The mask was then removed after the UVO treatment (Figure 6.22e) and the PDMS strip was released in both planar directions simultaneously, resulting in a concave microlens array (Figure 6.22f). Concave lenses were always obtained with prestrains of 10 to30% and UVO treatment of 15 to 60 min. A convex microlens array could be obtained by replica molding from the concave lenses in PDMS [19].

When a strain was applied onto the fabricated microlens array, the focal length could be tuned [19]. Figure 6.23 shows that the focal length of the concave microlenses increased rapidly when the strain approached prestrain. It should also be noted that the concave lenses had a far greater tuning range than the convex ones. The authors believed that this was to due to the fundamental difference between the fabrication processes of the two types of lenses. The curved structure of the concave microlenses was formed under buckling force when the prestrain was released. Thus, by design, the concave microlens could be completely flattened when the applied strain became equal to the prestrain.

Conversely, the convex lenses were replicated from the concave microlens and no buckling was involved. Hence the shape could never be flat, even under mechanical strain [19].

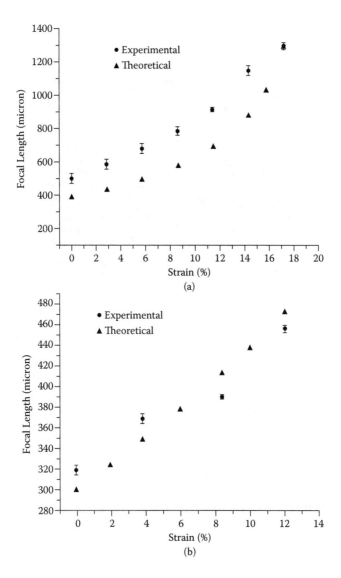

FIGURE 6.23
Focal length variation of (a) concave and (b) convex microlens arrays. (*Source:* Chandra, D., S. Yang, and P.C. Lin. 2007. *Applied Physics Letters*, 91(25). With permission.)

6.4 Colloidal Hydrogel Dynamically Tunable Microlenses

Lyon's group explored the application of hydrogels on dynamically tunable microlenses [20–22]. Poly (N-isopropylacrylamide) hydrogel materials can respond to external stimuli such as temperature, pH, ionic strength, electric field, and antibody–antigen interactions by swelling or shrinking transitions

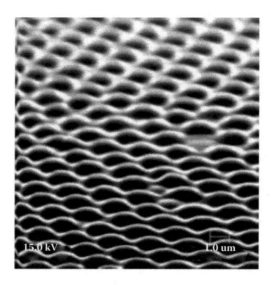

FIGURE 6.24
SEM image of hydrogel microlens array taken at grazing angle with respect to substrate. The microgels formed an ordered array of planoconvex shapes. (*Source:* Kim, J., M.J. Serpe, and L.A. Lyon. 2004. *Journal of the American Chemical Society*, 126(31), 9512–9513. With permission.)

[23–25]. These materials are stimuli-responsive, making them excellent candidates for use in dynamically tunable microlens structures.

Microlens arrays can be fabricated by assembling poly(N-isopropylacrylamide-co-acrylic acid) (pNIPAAm-AAc) microgels onto a 3-aminopropyltrimethoxysilane (APTMS)-functionalized glass substrate via an electrostatic self-assembly method [20,26]. Figure 6.24 shows an example of a microlens array fabricated in this approach. Planoconvex lens formation was due to the deformation of the microgels in an anisotropic fashion during substrate attachment because of the mechanical softness characteristics of microgels [22].

Figure 6.25 shows the focal length tunability of a single hydrogel microlens in response to pH at 25°C [22]. The differential interference contrast (DIC) images in Figure 6.25a and b indicate that the substrate-bound microgel was compact at pH 3.0 and swollen at pH 6.5. Since the diameter and curvature of the substrate-bound microgels were tunable in response to pH, the lens power was similarly tunable. The lens power tunability is illustrated in Figures 6.25c and d.

6.5 Liquid Microlenses Tuned by Environmental Stimuli-Responsive Hydrogels

6.5.1 Microlenses Actuated by Hydrogels

Stimuli-responsive hydrogels can be used directly as microlenses and also serve as actuators for microlenses. Jiang's group used stimuli-responsive

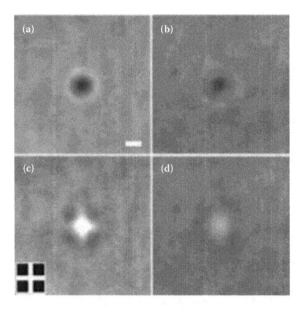

FIGURE 6.25
DIC microscopy images of substrate-bound microgel at pH 3.0 (a) and 6.5 (b) at 25°C. (c) and (d) Projection of cross-like pattern (inset) through microgel at respective pH values. Microgel at pH 3.0 brings into focus a cross-like pattern placed conjugate to the objective back focal plane, while the same microgel at pH 6.5 cannot focus the cross-like pattern at the same focal point. The scale bar represents 1 μm. (*Source:* Kim, J., M.J. Serpe, and L.A. Lyon. 2004. *Journal of the American Chemical Society,* 126(31), 9512–9513. With permission.)

hydrogels to sense changes in temperature [27,28] and pH [27,29] and control the shapes of liquid microlenses accordingly. The actuation of the hydrogels is reversible. This method takes advantage of the excellent properties of hydrogels as actuators, while allowing the use of other materials with better optical transparency than hydrogels, for example, water and oil.

In the first work reporting this approach, Dong et al. used a meniscus between water and oil as an optical lens and adjusted its focal length by changing its curvature [27]. Figure 6.26 shows the design and mechanism of the microlens. The basic design consists of a stimuli-responsive hydrogel ring placed within a microfluidic channel system and sandwiched between a glass plate and aperture slip, the latter with an opening centered over the ring. The microchannels were filled with water, and oil was placed on top of this structure and capped with a glass cover slip.

The sidewall and bottom surface of the aperture were hydrophilic and the top surface was hydrophobic to pin the water–oil meniscus along the hydrophobic–hydrophilic (H-H) boundary—the top edge of the aperture opening. When exposed to an appropriate stimulus, the hydrogel ring underneath responded by expanding or shrinking based on the absorption or release of water via the hydrogel network interstitials [25] This led to a change in the volume of the water droplet located in the middle of the ring.

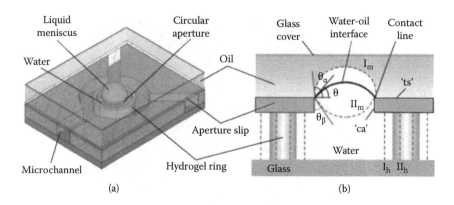

FIGURE 6.26

(a) Three-dimensional schematic of water–oil interface that forms a liquid microlens. Microchannels allow fluid flow to microlens structure. (b) Variable focus mechanism. Hydrophilic sidewall and bottom surface and hydrophobic top surface of the aperture pin the water–oil meniscus along the edge. Expansion and contraction of the hydrogel regulate shape of the liquid meniscus by changing the angle of the pinned water–oil interface to the edge. (*Source:* Dong, L., A.K. Agarwal, D.J. Beebe et al. 2006. *Nature*, 442(7102), 551–554. With permission.)

The net volumetric changes—the change in the volume enclosed by the ring and the change in water droplet volume—caused a change in the pressure difference P across the water–oil interface. P directly determined the geometry of the liquid meniscus. Because the meniscus was pinned at the H-H boundary and kept stationary, volume changes were translated into a change in the curvature of the meniscus, and hence the tuning of the focal length of the microlens.

The authors demonstrated microlenses responsive to temperature and pH values, using respective responsive hydrogels: a thermo-responsive hydrogel based on N-isopropylacrylamide (NIPAAm), and two pH-responsive hydrogels based on acrylic acid (AA) and 2-hydroxyethyl methacrylate (HEMA), and 2-(dimethylamino)ethyl methacrylate (DMAEMA) and HEMA, respectively [27,28].

Figure 6.27 shows how a temperature-responsive microlens adjusted its focal length in both positive and negative regimes as the temperature varied. A response time of 20 to 25 sec was observed [27]. Figure 6.28 shows the focal length variation of a pH-responsive liquid microlens with respect to the pH of its aqueous environment [27]. The response time of the pH-sensitive microlens was approximately 12 sec.

Besides using an enclosed elastic polymer ring made of stimuli-responsive hydrogel for the actuators, Dong et al. further implemented hydrogel-actuated liquid microlenses by harnessing the horizontal expansion and contraction of the hydrogel ring separating the lens liquid from its surrounding fluid in the microfluidic channel [29]. Figure 6.29 shows the mechanism of this type of hydrogel-driven liquid microlens.

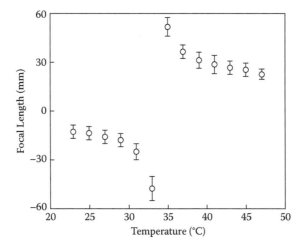

FIGURE 6.27
Focal length of temperature-responsive liquid microlens as function of temperature. Lens is divergent between 23.8°C (focal length = –11.7 mm) and 33.8°C (focal length approaches –∞). Between 33.8 and 47.8°C, the microlens becomes convergent with positive focal length from +∞ to 22.8 mm. (*Source:* Dong, L., A.K. Agarwal, D.J. Beebe et al. 2006. *Nature*, 442(7102), 551–554. With permission.)

A set of microposts made of pH-responsive hydrogel was constructed in a microfluidic chamber. A circular aperture was formed in a flexible polymer slip. The sidewall and top side of the aperture were treated to be hydrophilic and hydrophobic, respectively. Again, since aqueous solutions remained only on hydrophilic pathways at pressures below a critical value, part of the water-based liquid attached to the sidewall could form a liquid meniscus protruding downward at low pressures and upward at high pressures.

The water-based liquid containing stimuli acted as the lens liquid flowing in the chamber; oil was used to prevent the evaporation and also served as the lens liquid. The pinned water–oil interface again formed the liquid microlens. When the environmental pH changed, the hydrogel microposts could expand to bend a flexible aperture slip upward or restore its position at the contracted state. Correspondingly, the liquid meniscus could bow downward or be pressed to bulge out of the aperture. The shape of the microlens was changed in terms of the angle to the aperture slip; therefore the focal length of the microlens was tuned.

6.5.2 Microlenses Actuated through Nanoparticle-Incorporated Hydrogels

Zeng et al. extended the hydrogel-driven liquid microlens technology and presented a liquid variable focus microlens tuned by light-responsive hydrogels [30]. Figure 6.30 shows the schematics and optical images of a tunable

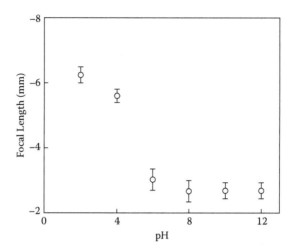

FIGURE 6.28
Focal length of pH-responsive liquid microlens using AA hydrogel as function of pH. (*Source:* Dong, L., A.K. Agarwal, D.J. Beebe et al. 2006. *Nature*, 442(7102), 551–554. With permission.)

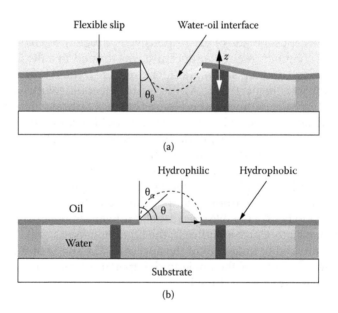

FIGURE 6.29
Formation of microlens at interface between oil and aqueous solution. Interface is pinned stably at a hydrophobic–hydrophilic boundary along circular aperture. Volumetric changes of hydrogel microposts cause flexible aperture slip to bend in the z direction. The pinned water–oil interface is pressed downward or upward, thus tuning the focal length. (*Source:* Dong, L. and H. Jiang. 2006. *Applied Physics Letters*, 89(21), 211120. With permission.)

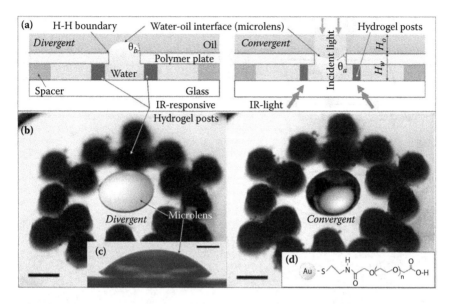

FIGURE 6.30
Schematic and optical images of IR light-actuated tunable liquid microlens. Structure is similar to that depicted in Figure 6.26. (a) Schematics of microlens when it is divergent and convergent, respectively. (b) Optical images of microlens with 18 hydrogel microposts in divergent and convergent states, respectively. (c) Side profile of water meniscus of microlens in divergent status at starting point. Scale bars represent 1 mm. (d) Chemical structure of gold nanoparticles coated with thiolated PEG ligands. (*Source: Zeng, X. and H. Jiang. 2008. Applied Physics Letters, 93(15), 151101. With permission.*)

liquid microlens actuated by an IR light-responsive hydrogel. The device had the same mechanism as the previous devices but used a modified hydrogel.

Multiple microposts made of IR light-responsive hydrogel were photopatterned in a water container to actuate the microlens under IR light irradiation. The IR light-responsive hydrogel consisted of the thermoresponsive NIPAAm hydrogel and water soluble gold nanoparticles with distinct and strong optical absorption of IR light and had high heat efficiency. With the IR light turned on, the gold nanoparticles absorbed the IR light, generating heat to cause the hydrogel to contract. When the IR light was turned off, the heat dissipated to the surrounding fluid and the hydrogel expanded back to its original volume. Such volumetric change could thus change the radius of curvature of a pinned water–oil interface as in previous devices. Hence, the tuning of the focal length of the liquid microlens formed via the water–oil interface was controlled by the IR light [30].

Figure 6.31 shows how the IR-responsive liquid microlens functioned when its focal length was tuned [30]. Figure 6.31a shows the scanning image planes using the microlens. Two objects were placed below the glass substrate at different distances. A CCD-coupled stereoscope was placed above the microlens to monitor and record the images. One scanning cycle included a forward

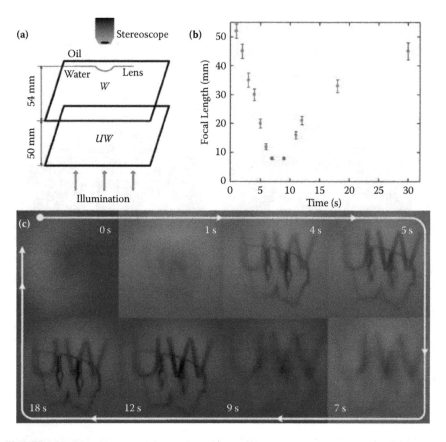

FIGURE 6.31
(a) Scanning image planes using IR-responsive liquid tunable microlens. W and UW are printed on transparency films (54 and 104 mm, respectively) below the glass substrate with microlens. A CCD-coupled stereoscope is placed above the microlens to monitor and record images. (b) Dynamic change in positive focal length of microlens (convergent) in one scanning cycle as function of time. (c) Frame sequence of focused images in one scanning cycle obtained by tuning microlens. (*Source:* Zeng, X. and H. Jiang. 2008. *Applied Physics Letters*, 93(15), 151101. With permission.)

scanning with IR light on and a reverse scanning with IR off. The dynamic change in the positive focal length of the microlens in one scanning cycle is plotted in Figure 6.31b. Figure 6.31c shows the frame sequence of the focused images obtained by tuning the microlens within one scanning cycle. During both the forward and reverse scanning cycles, the microlens was able to clearly image the two objects at different time instants with the right focal lengths.

6.5.3 Microlens Array Actuated through Thermoresponsive Hydrogels

The hydrogel-driven tunable liquid microlenses can be extended to microlens arrays in which microlenses can be individually and independently

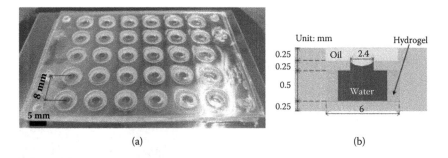

(a)   (b)

FIGURE 6.32
(a) Optical image of structure of microlens array without lens liquids. A array consists of 5 × 6 microlenses. (b) Cross-sectional schematic of microlens showing specifications of its size. (*Source:* Zeng, X.F., C.H. Li, D.F. Zhu et al. 2010. *Journal of Micromechanics and Microengineering,* 20(11), 115035. With permission.)

tuned in focal length. Zeng et al. demonstrated the feasibility of this concept as shown in Figure 6.32 [31]. Figure 6.32a shows an optical image of one 5 × 6 microlens array before filling the liquids. The image was taken by a camera from an oblique angle. Geometric parameters of each microlens are illustrated in Figure 32b.

Figure 6.33 shows the operation of the tunable liquid microlens array [31]. Each microlens in the array may be actuated individually by controlling local temperature. Resistive microheaters were placed under the microlens array and each microheater was connected to an individual temperature controller. Due to the individual tuning in focal length, the microlenses in the array could image objects independently.

6.5.4 Endoscopes Utilizing Tunable Microlenses

Endoscopes are important medical instruments that produce images of the interior surfaces of cavities in human bodies, for example the gastrointestinal (GI) tract including the esophagus, stomach, colon, and part of the small bowel; joint spaces; abdominal cavity; thorax; and other structures. In recent years, through technical improvements like wireless capsule endoscopy [32] and robotic surgery [33], minimally invasive surgery has become a common treatment.

Current fiber endoscopes use single opticals such as rod lenses and gradient index (GRIN) lenses at the distal ends of fibers or in front of charged-couple devices (CCDs) for imaging [34]. However, the focal length of these single lenses is not tunable, and it is difficult to manufacture and assemble a conventional zooming lens in an endoscope due to its small interior space. Hence endoscopic surgeons need extensive training because of the constant and highly complex manual maneuvering of endoscopes during procedures [35].

Tunable focus microlenses integrated at the ends of fiber endoscopes could scan the areas of interest with minimal or no back-and-forth movements of

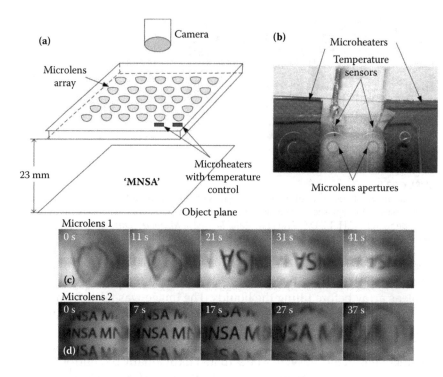

FIGURE 6.33
(a) Set for forming images on camera through tunable liquid microlens array. Object plane (MNSA) was placed 23 mm below microlens array. Camera-coupled stereomicroscope monitored and recorded microlens images. Two resistive microheaters were placed under microlenses, each connected to an individual temperature controller. (b) Image of two microlenses with corresponding microheater and temperature sensor. (c) Frame sequence of images obtained from microlens 1. Images from microlens 1 were inverted, enlarged, and real. At 21 sec, image was clear and sharp. (d) Frame sequence of images obtained from microlens 2. Images were erect, enlarged, and virtual. At 17 sec, image was clear and sharp. (*Source:* Zeng, X.F., C.H. Li, D.F. Zhu et al. 2010. *Journal of Micromechanics and Microengineering*, 20(11), 115035. With permission.)

the scope instruments. The variable focal lengths of microlenses allow different depths of focus and thus spatial depth perception in images is possible.

Based on results with liquid microlenses actuated through IR light-responsive hydrogels described in the previous section, Jiang's group demonstrated prototype endoscopes with tunable focus liquid microlenses integrated at their distal ends actuated through IR light [36,37]. Two types of endoscopes were presented [36]. Hydrogel microstructures were defined by fibers in Endoscope I and by masks in Endoscope II. Table 6.3 lists the parameters of both types of endoscopes.

Figure 6.34 shows a prototype endoscope with a liquid tunable focus microlens integrated at the distal end [36]. Two sets—actuation optical fibers and an image acquisition fiber bundle—were connected to the back side of

TABLE 6.3

Parameters of Two Types of Endoscopes

Parameter	Endoscope I	Endoscope II
Core diameter of actuation optical fiber (μm)	600	1000
Number of actuation optical fibers	6	12
Method of defining hydrogel microstructures	Via optical fibers also used for hydrogel actuation	Via masks
Number of defined hydrogel posts	6	48
Blocking layer	No	Yes
Diameter of image acquisition fiber bundle (mm)	6	2
Resolution of image acquisition fiber bundle	3,200	17,000
Camera connected to image acquisition fiber bundle	Regular camera	Industrial camera

Source: X. Zeng, C.T. Smith, J.C. Gould et al. 2011. *Journal of Microelectromechanical Systems,* 20(3), 583–593. With permission.

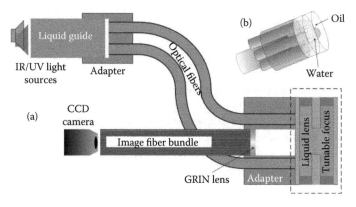

FIGURE 6.34

(a) Schematic and (b) three-dimensional view of prototype endoscope with liquid tunable focus microlens integrated at the distal end and actuated by infrared light. Two sets of optical fibers are used: actuation fibers and image acquisition fiber bundle. Infrared light is transmitted via actuation optical fibers to tune focal length of microlens at end of fibers. Images from microlens are transferred to a camera by image acquisition fiber bundle. (*Source:* X. Zeng, C.T. Smith, J.C. Gould et al. 2011. *Journal of Microelectromechanical Systems,* 20(3), 583–593. With permission.)

the microlens. Infrared (IR) light was transmitted via the actuation optical fibers to actuate hydrogel microstructures and thus vary the focal length of the microlens at the end of the fibers. Images from the microlens were transferred to a camera via the image acquisition fiber bundle.

Zeng et al. [36] also demonstrated that the reported endoscopes could be used to obtain clinically relevant images. An endoscopic environment was constructed to simulate the human colon and observed by Endoscope II.

FIGURE 6.35
Frame sequence of focused images of simulated human colon and polyps obtained by Endoscope II in one scanning cycle of microlens. Initially, three polyps could be observed in the field. From 1.8 sec, infrared light was on and angle of view (AOV) began to increase. At 4.6 sec, one extra polyp at the other end of the simulated colon lumen could be observed. At 5.2 sec, the light was off and AOV of the microlens began to decrease. At 11.4 sec, the microlens returned to original status. (*Source:* X. Zeng, C.T. Smith, J.C. Gould et al. 2011. *Journal of Microelectromechanical Systems*, 20(3), 583–593. With permission.)

Simulated tissue was used to represent colonic polyps with diameters ranging from 3 to 10 mm, mimicking a clinically relevant range of polyps [38]. Figure 6.35 depicts the frame sequence of the focused images obtained from Endoscope II in one scanning cycle.

6.5.5 Microlens Arrays on Flexible and Curved Surfaces

Zhu et al. devised responsive hydrogel-driven microlenses fabricated on non-planar surfaces [39,40]. Tunable liquid microlenses were fabricated onto flexible polymer substrates and wrapped onto a dome-like structure. This work was inspired by the compound eyes of insects found in nature. Figure 6.36 is a photograph of the compound eyes of a dragonfly. A compound eye consists of many small eyes (ommatidia) containing small individual lenses. The ommatidia are arranged on a spherical structure. A conspicuous advantage of this design is the large FOV. However, the focal lengths of ommatidia thus have lower resolution.

The work of Zhu et al. is intriguing in that it combines the benefits of tunable liquid microlenses and the large FOV offered by compound eyes. Figure 6.37 illustrates a tunable liquid microlens array with six elements fabricated on a hemisphere. The fabrication of these arrays was described in Example 3 in Chapter 3. The microlenses were made in individual islands connected via thin polymer bridges so that the stress from wrapping the whole structure onto the

FIGURE 6.36
Compound eyes of dragonfly. (Photo courtesy of Christopher J. Murphy.)

FIGURE 6.37
Tunable liquid microlens array fabricated on hemisphere. Lenses are individually tunable in their focal lengths. (*Source:* Difeng, Z., Z. Xuefeng, L. Chenhui et al. 2011. *Journal of Microelectromechanical Systems*, 20(2), 389–395. With permission.)

non-planar surface mostly appeared in the bridge structures rather than the islands. This was critical to ensure that the microlenses would not experience much stress and distortion that could severely affect their optical performance.

6.6 Oscillating Microlens Arrays Driven by Sound Waves

Lopez et al. introduced harmonically driven liquid lenses with oscillating focal lengths that are particularly suited for fast focal length changes [41]. The liquid lenses were constructed by coupling two droplets through a cylindrical hole. The opposing curvature of the droplets created a spring-like force that made the system act as an oscillator (Figure 6.38). Forcing the system at resonance allowed the oscillatory motion to be sustained with little energy input.

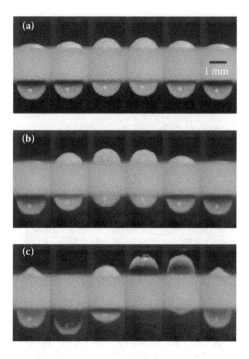

FIGURE 6.38
Time series of pinned contact oscillating liquid lens driven by external pressure at resonance but with different forcing amplitudes. Driving amplitudes are 5.5 Pa (a), 23.4 Pa (b), and 31.7 Pa (c), respectively. Liquid volume is where R is lens radius. Lens is driven at 49 Hz. Time between frames is 4 msec. Scale bar is 1 mm. (*Source:* Lopez, C.A. and A.H. Hirsa. 2008. *Nature Photonics,* 2(10), 610–613. With permission.)

The liquid-to-solid contact lines were pinned by a hydrophobic substrate; this significantly reduced energy dissipation from movements of the contact lines and viscosity. The natural frequency of the system scaled with the radius of the lens as $R^{-3/2}$, and the resulting very high frequency response was obtained with a modestly sized lens [41].

The focal lengths of such oscillatory liquid microlenses can be predicted by using the lumped mass model for spherical caps [41]. Figure 6.39 shows measurements and predicted motion of the center of mass of a liquid lens. The focal length was defined based on the distance from a droplet apex. The focal length was two to three times the radius of the droplet and the response time was 20 msec. Higher frequencies could be achieved at smaller scales; for a given scale, the maximum frequency occurred at volumes of liquid droplets approaching 0. Furthermore, as the length scale decreased, the effect of gravity diminished, resulting in the appearance of a low frequency cusp [41].

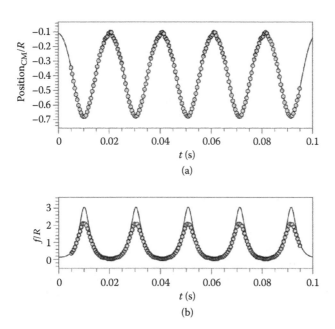

FIGURE 6.39
Predicted and measured motion of oscillatory liquid microlens along with focal length. (a) Motion of center of mass for lens with liquid volume of forced at frequency of 49 Hz and pressure of 8.0 Pa. (b) Thick line shows focal length computed using predicted position of center of mass. (*Source:* Lopez, C.A. and A.H. Hirsa. 2008. *Nature Photonics*, 2(10), 610–613. With permission.)

References

1. S. H. Ahn and Y. K. Kim, "Proposal of human eye's crystalline lens-like variable focusing lens," *Sensors and Actuators A - Physical*, vol. 78, pp. 48–53, Nov 1999.
2. D. Y. Zhang, V. Lien, Y. Berdichevsky, J. Choi, and Y. H. Lo, "Fluidic adaptive lens with high focal length tunability," *Applied Physics Letters*, vol. 82, pp. 3171–3172, May 2003.
3. A. Werber and H. Zappe, "Tunable microfluidic microlenses," *Applied Optics*, vol. 44, pp. 3238–3245, 2005.
4. N. Chronis, G. L. Liu, K. H. Jeong, and L. P. Lee, "Tunable liquid-filled microlens array integrated with microfluidic network," *Optics Express*, vol. 11, pp. 2370–2378, Sep 2003.
5. K. H. Jeong, G. L. Liu, N. Chronis, and L. P. Lee, "Tunable microdoublet lens array," *Optics Express*, vol. 12, pp. 2494–2500, May 2004.
6. J. Chen, W. S. Wang, J. Fang, and K. Varahramyan, "Variable-focusing microlens with microfluidic chip," *Journal of Micromechanics and Microengineering*, vol. 14, pp. 675–680, May 2004.

7. G. H. Feng and Y. C. Chou, "Fabrication and characterization of optofluidic flexible meniscus-biconvex lens system," *Sensors and Actuators A-Physical*, vol. 156, pp. 342–349, Dec 2009.

8. M. Agarwal, R. A. Gunasekaran, P. Coane, and K. Varahramyan, "Polymer-based variable focal length microlens system," *Journal of Micromechanics and Microengineering*, vol. 14, pp. 1665–1673, Dec 2004.

9. P. M. Moran, S. Dharmatilleke, A. H. Khaw, K. W. Tan, M. L. Chan, and I. Rodriguez, "Fluidic lenses with variable focal length," *Applied Physics Letters*, vol. 88, p. 041120, 2006.

10. H. B. Yu, G. Y. Zhou, F. K. Chau, F. W. Lee, S. H. Wang, and H. M. Leung, "A liquid-filled tunable double-focus microlens," *Optics Express*, vol. 17, pp. 4782–4790, Mar 2009.

11. G. R. Xiong, Y. H. Han, C. Sun, L. G. Sun, G. Z. Han, and Z. Z. Gu, "Liquid microlens with tunable focal length and light transmission," *Applied Physics Letters*, vol. 92, p. 241119, Jun 2008.

12. D. Y. Zhang, N. Justis, and Y. H. Lo, "Fluidic adaptive lens of transformable lens type," *Applied Physics Letters*, vol. 84, pp. 4194–4196, May 2004.

13. L. Pang, U. Levy, K. Campbell, A. Groisman, and Y. Fainman, "Set of two orthogonal adaptive cylindrical lenses in a monolith elastomer device," *Optics Express*, vol. 13, pp. 9003–9013, Oct 2005.

14. H. Ren and S. T. Wu, "Variable-focus liquid lens by changing aperture," *Applied Physics Letters*, vol. 86, p. 211107, May 2005.

15. H. Ren, D. Fox, P. A. Anderson, B. Wu, and S.-T. Wu, "Tunable-focus liquid lens controlled using a servo motor," *Optics Express*, vol. 14, pp. 8031–8036, 2006.

16. H. W. Ren and S. T. Wu, "Variable-focus liquid lens," *Optics Express*, vol. 15, pp. 5931–5936, May 2007.

17. D. P. Holmes and A. J. Crosby, "Snapping surfaces," *Advanced Materials*, vol. 19, pp. 3589–3593, Nov 2007.

18. Y. Forterre, J. M. Skotheim, J. Dumais, and L. Mahadevan, "How the Venus flytrap snaps," *Nature*, vol. 433, pp. 421–425, Jan 2005.

19. D. Chandra, S. Yang, and P. C. Lin, "Strain responsive concave and convex microlens arrays," *Applied Physics Letters*, vol. 91, p. 251912, Dec 2007.

20. M. J. Serpe, J. Kim, and L. A. Lyon, "Colloidal hydrogel microlenses," *Advanced Materials*, vol. 16, pp. 184–187, Jan 2004.

21. C. D. Jones, M. J. Serpe, L. Schroeder, and L. A. Lyon, "Microlens formation in microgel/gold colloid composite materials via photothermal patterning," *Journal of the American Chemical Society*, vol. 125, pp. 5292–5293, May 2003.

22. J. Kim, M. J. Serpe, and L. A. Lyon, "Hydrogel microparticles as dynamically tunable microlenses," *Journal of the American Chemical Society*, vol. 126, pp. 9512–9513, Aug 2004.

23. H. G. Schild, "Poly (N-Isopropylacrylamide): experiment, theory and application," *Progress in Polymer Science*, vol. 17, pp. 163–249, 1992.

24. J. H. Holtz and S. A. Asher, "Polymerized colloidal crystal hydrogel films as intelligent chemical sensing materials," *Nature*, vol. 389, pp. 829–832, Oct 1997.

25. Y. Osada, J. P. Gong, and Y. Tanaka, "Polymer gels (Reprinted from Functional Monomers and Polymers, pg 497–528, 1997)," *Journal of Macromolecular Science-Polymer Reviews*, vol. C44, pp. 87–112, Feb 2004.

26. D. Gan and L. A. Lyon, "Interfacial nonradiative energy transfer in responsive core-shell hydrogel nanoparticles," *J. Am. Chem. Soc.*, vol. 123, pp. 8203–8209, Aug 2001.

27. L. Dong, A. K. Agarwal, D. J. Beebe, and H. Jiang, "Adaptive liquid microlenses activated by stimuli-responsive hydrogels," *Nature*, vol. 442, pp. 551–554, Aug 2006.

28. L. Dong, A. K. Agarwal, D. J. Beebe, and H. Jiang, "Variable-focus liquid microlenses and microlens arrays actuated by thermoresponsive hydrogels," *Advanced Materials*, vol. 19, pp. 401–405, Feb 2007.

29. L. Dong and H. Jiang, "pH-adaptive microlenses using pinned liquid-liquid interfaces actuated by pH-responsive hydrogel," *Applied Physics Letters*, vol. 89, p. 211120, Nov 2006.

30. X. Zeng and H. Jiang, "Tunable liquid microlens actuated by infrared light-responsive hydrogel," *Applied Physics Letters*, vol. 93, p. 151101, Oct 2008.

31. X. Zeng, C. Li, D. Zhu, H. J. Cho, and H. Jiang, "Tunable microlens arrays actuated by various thermo-responsive hydrogel structures," *Journal of Micromechanics and Microengineering*, vol. 20, p. 115035, Nov 2010.

32. A. Moglia, A. Menciassi, M. O. Schurr, and P. Dario, "Wireless capsule endoscopy: from diagnostic devices to multipurpose robotic systems," *Biomedical Microdevices*, vol. 9, pp. 235–243, Apr 2007.

33. D. Oleynikov, "Robotic Surgery," *Surgical Clinics of North America*, vol. 88, pp. 1121–1130, Oct 2008.

34. J. Knittel, L. Schnieder, G. Buess, B. Messerschmidt, and T. Possner, "Endoscope-compatible confocal microscope using a gradient index-lens system," *Optics Communications*, vol. 188, pp. 267–273, Feb 2001.

35. P. B. Cotton, C. B. Williams, R. H. Hawes, and B. P. Saunders, *Practical Gastrointestinal Endoscopy: the Fundamentals*, 6th edn. ed. West Sussex: Willey-Blackwell, 2008.

36. X. Zeng, C. T. Smith, J. C. Gould, C. P. Heise, and H. Jiang, "Fiber endoscopes utilizing liquid tunable-focus microlenses actuated through infrared light," *Journal of Microelectromechanical Systems*, vol. 20, pp. 583–593, 2011.

37. X. Zeng and H. Jiang, "An endoscope utilizing tunable-focus microlenses actuated through infrared light," in *The 15th International Conference on Solid-State Sensors, Actuators and Microsystems* Denver, CO, USA, 2009, pp. 1214–1217.

38. M. L. Corman, *Colon and Rectal Surgery*, 5th edn ed. Philadelphia: Lippincott Williams & Wilkins, 2005.

39. D. Zhu, C. Li, X. Zeng, and H. Jiang, "Tunable-focus microlens arrays on curved surfaces," *Applied Physics Letters*, vol. 96, p. 081111, Feb 2010.

40. D. Zhu, X. Zeng, C. Li, and H. Jiang, "Focus-tunable microlens arrays fabricated on spherical surfaces," *Journal of Microelectromechanical Systems*, vol. 20, pp. 389–395, 2011.

41. C. A. Lopez and A. H. Hirsa, "Fast focusing using a pinned-contact oscillating liquid lens," *Nature Photonics*, vol. 2, pp. 610–613, Oct 2008.

7

Horizontal Microlenses
Integrated in Microfluidics

In previous chapters, many emerging microlenses based on various mechanisms covering non-tunable and tunable types were presented. However, these microlenses have their optical axes perpendicular to the substrates, thus requiring optical alignment of the different layers. This causes complicated structures for applications such as labs on chips. In this chapter, we discuss horizontal microlenses integrated in microfluidics. Their optical axes are parallel to the substrates of the microfluidic networks. These horizontal microlenses include those formed by a hot embossing process; hydrodynamically tuned cylindrical microlenses; and tunable and movable liquid droplets as microlenses.

7.1 Introduction

A horizontal microlens is based on observation and light activation from the top. Other requirements involve light control and manipulation from the sides of microfluidic channels. Several researchers have dealt with the fabrication of embedded waveguides using various materials and technologies [1–7]. One key question is how to guide light into the waveguides. Various designs integrating fibers into microchips have been proposed [3,6,8]. The key components in these designs are in-plane microlenses.

In the previous chapters, the optical axes of the tunable microlenses discussed were perpendicular to their substrates. However, in-plane microlenses utilizing the curvature formed by the interfaces between two immiscible liquids or between air and liquid can be introduced to observe objects in microfluidic channels. Their optical axes are parallel to the substrate of microlenses—a distinctly different system from what we have seen previously.

7.2 Two-Dimensional (2D) Microlenses

Camou et al. proposed a design including optical 2D microlenses in a PDMS layer [9]. The principle of the PDMS lenses can be easily understood from Figure 7.1. An optical fiber is inserted into a channel ending with a curved interface. The two media on each side of this interface (air and PDMS) possess different refractive indices so that the light beam going through this interface will be bent and focused according to the curvature of radius of the interface and the incident angle of the light beam.

Figure 7.2 shows the images of the mold and the PDMS microlens fabricated from it. It should be noted that this design cannot modify the light beam along the Z direction. The efficiency of these lenses is then limited to a 2D plane defined by both the X axis and Y axis parallel to the glass substrate—the PDMS layer interface.

Hsiung et al. did not use a PDMS layer. They utilized a hot embossing process to form microlenses [10]. The hot embossing technique is relatively simple and reliable [11]. The manufacturing process involved three major steps: (1) fabrication of the glass-based master mold, (2) fabrication and bonding of the polymethylmethacrylate (PMMA) chips, and (3) insertion of the optic fibers.

Hsiung et al. employed polished soda-lime glass substrates to fabricate the master mold of the microfluidic device [10]. A simple and effective glass etching process was developed to form the microstructures. Briefly, the master mold for hot embossing was fabricated on the surface of a 3 mm thick soda-lime glass substrate. Initially, a thin chrome (Cr) film 1000 Å in thickness was deposited on the surface of the substrate. A thin layer of AZ4620 positive photoresist (PR) was then spin coated on the substrate to form an etching mask to define the pattern of the underlying chrome layer.

The patterned Cr layer in turn served as the etching mask for the subsequent glass etching. Inverse structures with heights of 40 μm were then formed in a commercially available etchant. After the glass-based master mold was fabricated, a hot-embossing technique was conducted at 130°C for 5 min to transfer the inverse images of the microchannels and microfocusing lenses onto the PMMA substrates.

This hot-embossing process simultaneously defined the two pairs of optic fiber channels. After a demolding process, the microstructures with depths of 40 μm were formed. Fluid via holes were then drilled through an upper PMMA substrate. The upper and lower symmetrical PMMA substrates were carefully aligned under a microscope and their edges prebonded with a UV-sensitive glue. After alignment and the subsequent thermal bonding process, the two hot-embossed PMMA substrates formed a sealed microchannel.

Figure 7.3a is a photograph of the assembled microchip device. Two pairs of multimode optic fibers were integrated with the plastic chip to facilitate multiple wavelength detection of two different types of fluorescence dyes.

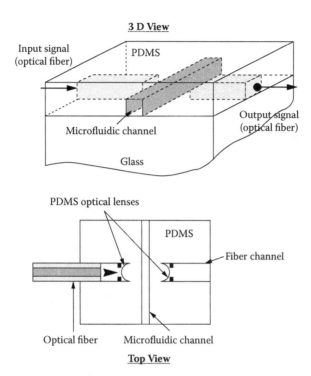

FIGURE 7.1
Experimental setup for 2D microlenses. (*Source:* Camou, S., H. Fujita, and T. Fujii. 2003. *Lab on a Chip*, 3(1), 40–45. With permission.)

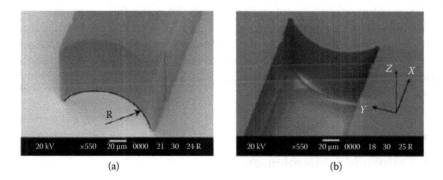

FIGURE 7.2
Characterization of device structures using scanning electron microscope. (a) SU8 mold with curved end serves as mold for 2D optical microlens. (b) PDMS layer with 2D optical microlens at end of the channel. (*Source:* Camou, S., H. Fujita, and T. Fujii. 2003. *Lab on a Chip*, 3(1), 40–45. With permission.)

Figure 7.3b is a close-up view of the focusing lens within the integrated chip. To provide a focused light source, thereby increasing the amplitude of the signal and enhancing detection performance, the focusing lens was

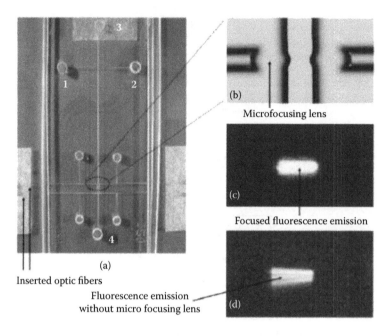

(a)

Inserted optic fibers

Fluorescence emission
without micro focusing lens

FIGURE 7.3
(a) Assembled microchip with microfocusing lenses integrated with microfluidics. Device measures 3 × 8 cm. (b) Magnified view of integrated microfocusing lens and embedded optic fiber. (c) Image of fluorescent dye in sample flow channel excited by laser source with focusing lens. (d) Image of fluorescent dye in sample flow channel excited by laser source without focusing lens. (*Source:* Suz-Kai Hsiung, C.H. 2005. *Electrophoresis*, 26(6), 1122–1129. With permission.)

positioned between the front end of the optic fiber and the sample flow channel. The light source was provided by a 50 mW argon laser (488 nm wavelength). Figures 7.3c and d show the resulting excitation of the fluorescent dye with and without the focusing microlens. The microfocusing lens structure enhances the emitted fluorescence signal coupled into the optic fiber of the detection system by 1.67 times.

7.3 Optofluidic Microlenses

Optofluidics is a new specialty that emerged in recent years. It benefits from the small scale of microfluidics and utilizes single or multiple fluidic flows to manipulate, guide, or control light at microscale and even at nanoscale. Many investigators have devised microlenses based on the optofluidic approach.

7.3.1 Tunable and Movable Liquid Microlenses

Dong et al. presented a tunable and movable liquid microlens in situ fabricated through a relatively simple process by manipulating fluids within a microfluidic network [12]. A de-ionized (DI) water droplet segmented by air was generated and guided to a T-shaped junction of microchannels by pneumatic fluid manipulation. After surface chemistry treatment, the edges at the corners of this junction achieved high surface energy and were able to obstruct the movement of the droplet.

When the air pressure difference applied to the droplet equaled the internal capillary pressure caused by the difference in the curvature between the two liquid–air interfaces of the droplet, the liquid–air interface at the junction could protrude from the microchannel and be steadily pinned along the edges. The shape of the other liquid–air interface, on the other hand, depended on the static contact angle of the liquid on the channel material under homogeneous pneumatic pressure. A liquid microlens was thus formed from the droplet.

Because of the fluidic nature of the liquid and the surface tension as a dominant force at microscale, varying the air pressure difference within a certain range could change the shape of the pinned liquid–air interface, thus tuning the focal length of the resulting microlens. With proper pneumatic controls, this microlens could be moved further within the microchannel at various predefined spots corresponding to individual junctions.

A typical design of a microchannel network would consist of a main fluidic channel, two air inlets with pneumatic pressure controls for the handling of fluids, and a lens channel, as shown in Figure 7.4. First, the authors introduced a DI water stream into the main channel at an infusion rate of 0.5 mL/min. When the stream passed junction J1, a syringe air pump S1 injected an air plug (16.2 µl volume; 1.013×10^5 Pa pressure; 0.9 mL/min infusion rate)

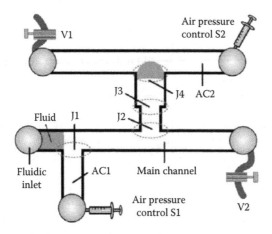

FIGURE 7.4
Microfluidic setup for in situ fabrication of liquid droplet-based microlens. To realize the movement of microlens from one junction to another, a small step is produced at junction J3. (*Source:* Dong, L. and H. Jiang. 2007. *Applied Physics Letters*, 91(4), 041109. With permission.)

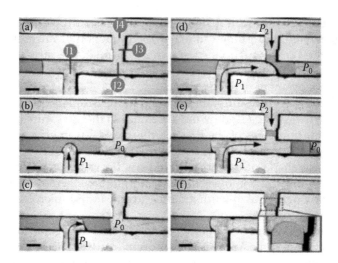

FIGURE 7.5
In situ fabrication of liquid droplet-based microlens within microchannel. (a) Stream is flowed into main channel. (b) and (c) Water droplet is cut from main stream by air pressure P_1 at junction J1. (d) and (e) Lens droplet is split into lens channel at junction J2. (f) Lens droplet stops at junction J3. P_0 = atmospheric pressure. Scale bars = 1 mm. (*Source:* Dong, L. and H. Jiang. 2007. *Applied Physics Letters,* 91(4), 041109. With permission.)

into air conduit AC1 to separate a water droplet from the main stream. This segmented water continued to advance.

When it arrived at junction J2, a lens droplet was split into the lens channel. The size of the lens droplet could be controlled by air pressures P_1 and P_2 from air conduits AC1 and AC2, respectively. Syringe air pump S2 dispensed another air plug (volume: 10.5 µl; pressure: 1.013×10^5 Pa; infusion rate: 0.9 mL/min) into AC2 to adjust P_2. This lens droplet subsequently advanced in the lens channel, and then stopped at the edges of the corners of J3 with a pressure difference $\Delta P = P_1 - P_2$ (~110 Pa in this example) over the droplet.

To sustain the microlens at these two edges, ΔP needed to be less than a critical pressure difference ΔP_C. During the formation of the microlens, valve V1 remained closed and V2 was opened to ambient air before a water droplet in the main channel arrived at it. The microlens could be removed by opening V1 and exerting air pressure to squeeze the droplet into the reservoir. The microlens also could be reformed on demand following the same procedures.

This approach to microlens production offers much flexibility and reconfigurability in operation. The authors also demonstrated that the *in situ* fabricated liquid microlens could be moved from one junction to another within the lens channel and be tuned in focal length, as shown in Figure 7.6.

When applying a ΔP larger than the critical pressure difference at junction J3, the microlens left J3. The lens droplet was subsequently stopped at J4 by immediately decreasing ΔP to less than the critical pressure difference

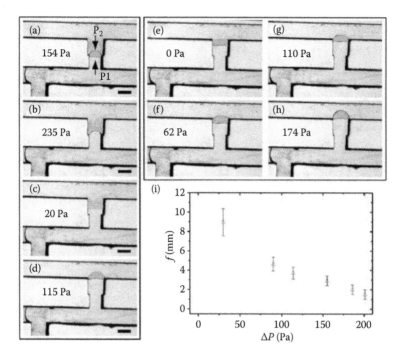

FIGURE 7.6
Reconfiguring in situ fabricated liquid microlens. The data in each snapshot show $\Delta P = P_1 - P_2$ across the microlens. (a) through (d) Repositioning from one junction to another. (e) through (h) Tuning focal length. (i) Focal length at various ΔP levels. Scale bars = 1 mm. (*Source:* Dong, L. and H. Jiang. 2007. *Applied Physics Letters*, 91(4), 041109. With permission.)

at J4. The microlens at junctions J3 and J4 could be tuned while being stably pinned along the edges at the corners of the specific junction. At a ΔP exceeding ΔP_{J4}, the microlens broke the geometrical obstruction at J4 and flowed into a reservoir through AC2 and V1. Figure 7.6i demonstrates that the focal length could be pneumatically controlled, varying from 1.5 to 8.9 mm over ΔP from 30 to 201 Pa.

7.3.2 Hydrodynamically Tunable Optofluidic Microlenses

Mao et al. demonstrated a hydrodynamically tunable optofluidic microlens allowing variable focusing of light within a microfluidic device [13]. The mechanism of the optofluidic tunable microlens is shown in Figure 7.7. The microlens was constructed from two fluids with different refractive indices, 5 M CaCl$_2$ solution ($n = 1.445$) and DI water ($n = 1.335$). Both fluids were injected into a microfluidic channel with a 90 degree curve. The adjacent injection of the two miscible fluids resulted in an optically smooth, nearly vertical interface, due to the laminar flow that dominated in microfluidic channels.

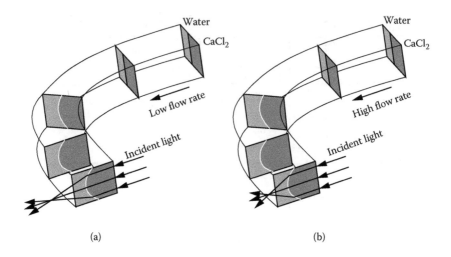

(a)   (b)

FIGURE 7.7
Mechanism of hydrodynamically tunable optofluidic cylindrical microlens. $CaCl_2$ solution bows outward into water due to centrifugal effect induced in curve. Shorter focal length is obtained after flow transitions from (a) low flow rate to (b) high flow rate. (*Source:* Mao, X.L., J.R. Waldeisen, B.K. Juluri et al. 2007. *Lab on a Chip*, 7(10), 1303–1308. With permission.)

Upon entering the curve, the fluid experienced a centrifugal force along the curved trajectory. The fluid flowing in the middle of the channel (where the flow velocity was the highest) experienced a higher centrifugal force than the surrounding flow. As a result, a pair of secondary counter-rotating vortices located in the upper and lower halves of the cross-sectional plane of the channel was induced.

The secondary vortical flow perturbed the fluidic interface, pulling the fluid in the middle of channel toward the outer channel wall and sweeping the fluid at the top and bottom of the channel toward the inner channel wall. Consequently, the originally flat fluidic interface bowed outward, creating a cylindrical microlens. The magnitude of the overall centrifugal effect (called interface bowing) was determined by the ratio of the inertial and centrifugal force to the viscous force. Therefore, the shape of the fluidic interface and the optical characteristic of the cylindrical microlens may be altered conveniently by changing the flow rate.

The authors performed quantitative analysis of the focused light intensity for the microlens device [13]. Normalized light intensity was measured by a detection optical fiber for flow rates ranging from 0 to 400 µl/min at intervals of 50 µl/min, as shown in Figure 7.8a. As the flow rate increased from 0 to 250 µl/min, the light focused into the detection fiber intensified as a result of the increasing microlens curvature and the focal plane approaching the aperture of the detection fiber (shorter focal length).

A linear correlation could also be extrapolated between the detection intensity and the flow rates between 50 and 200 µl/min. Light intensity

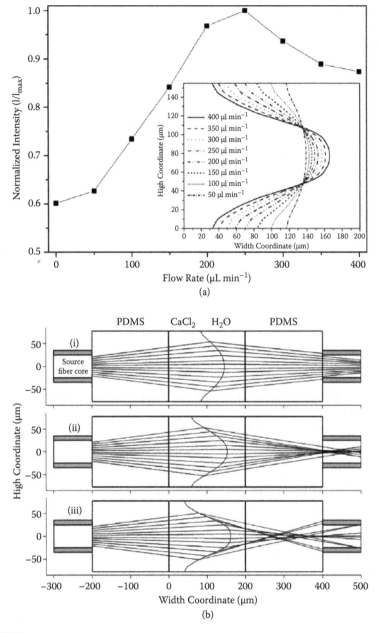

FIGURE 7.8
(a) Normalized light intensity from hydrodynamically tuned microlens, measured for flow rates ranging from 0 to 400 μl/min at 50 μl/min intervals. Inset: simulated results of fluidic interface at respective flow rates (except for 0 μl/min). (b) Ray tracing simulation reveals three focusing patterns: (i) under-focused mode at flow rate of 150 μl/min, (ii) well-focused mode at flow rate of 250 μl/min, and (iii) over-focused mode at flow rate of 350 μl/min. (*Source:* Mao, X.L., J.R. Waldeisen, B.K. Juluri et al. 2007. *Lab on a Chip,* 7(10), 1303–1308. With permission.)

subsequently decreased after reaching the maximum intensity at a flow rate of 250 μl/min. The reason was that the light was focused before reaching the aperture of the detection fiber. Simulation results of the microlens profile for flow rates ranging from 50 to 400 μl/min at 50 μl/min intervals are shown in the inset of Figure 7.8a.

Using simulated lens profiles, the authors performed a ray tracing simulation to calculate the focal length of the optofluidic microlens at different flow rates (Figure 7.8b) [13]. A cone-shaped divergent light source was employed to simulate the input beam of the fiber. Three different flow rates (150, 250, and 350 μl/min) resulted in three focusing patterns: (i) under-focused, (ii) well-focused, and (iii) over-focused. From (i) to (ii), the increase of the flow rate resulted in a decrease of focal length, causing the focal plane to approach the aperture of the detection fiber. The increased focusing power of the lens caused more of the divergent incident beams to bend into the detection fiber.

At a flow rate of 250 μl/min, most of the incident light beams were well focused near the fiber aperture (ii), corresponding to the maximum intensity observed in the experiment. From (ii) to (iii), over-focusing of the light caused the incident light beams to focus before reaching the fiber aperture and subsequently diverge. As a result, less light was coupled into the optical fiber, resulting in a decrease in the detection intensity.

7.3.3 Liquid Core and Liquid Cladding Microlenses

Tang et al. [14] and Seow et al. [15] described the design and operation of another type of dynamically reconfigurable microfluidic lens for use in microfluidic networks. The lens was formed by laminar flow of three streams of fluids. The refractive index of the central (core) stream was higher than the indices of the sandwiching (cladding) streams. The streams entered a microchannel having a region in which the channel expanded laterally.

Figure 7.9 shows the working principle of the formation of this type of liquid microlens with different lens curvatures in the expansion chamber achieved by tuning the flow rates of the three liquid streams. The configuration of the liquid lens was controlled hydrodynamically through the flow rates of the three streams. The central, core stream had a flow rate of V_{co}. The other two (left and right) cladding streams had flow rates of V_{cll} and V_{clr}, respectively.

If both cladding liquid streams had the same flow rate ($V_{cll} = V_{clr}$) and the core liquid stream had a V_{co} higher than that, a fluidic biconvex lens was formed. Note the two radii of curvature of the lens could be tuned individually with V_{cll}, V_{clr} and V_{co}. The radii of curvature increased when V_{cll} or V_{clr} increased. For example, when V_{clr} increased and was higher than V_{co}, the radius of curvature of the right interface increased. When the radius of curvature from the right side reached infinity, a planar convex lens was formed. When V_{clr} was further increased, a concave convex lens was formed [15].

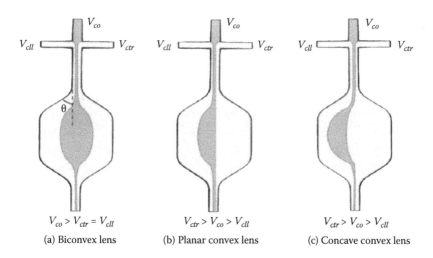

FIGURE 7.9
Formation of tunable liquid microlens dynamically tuned by flow rates of three streams. (*Source:* Seow, Y.C., A.Q. Liu, L.K. Chin et al. 2008. *Applied Physics Letters*, 93(8), 0841013. With permission.)

Figure 7.10 shows the shapes of such a lens under different flow conditions. The flow was considered quasi-2D since the height of the expansion chamber was much smaller than its width and length. Figure 7.10a shows the curvature ($R_{curvature}$) of the left and right interfaces of this biconvex lens as a function of the flow rates of the streams. Figure 7.10b shows the shapes of the lens obtained.

By varying the flow rates, the curvatures of the left and right interfaces were varied to obtain an extensive range of lens shapes: meniscus, planoconvex, and biconvex [14].

Figure 7.11 shows the focal distance of a liquid core and liquid cladding lens as a function of the flow rates of the streams [15]. Due to the focusing effect of the lens, the intensity of a laser beam passing through the lens significantly increased.

Unlike the designs with rectangular lens expanding chambers discussed above, Song et al. proposed a design with a circular chamber to realize a perfect lens profile [16]. They presented a mathematical model to describe the relationship between the lens shape and flow rate. Their circular chamber is shown in Figure 7.12 [16]. The relationship between the flow rate ratio and focal length of the symmetrical double convex optofluidic lens was:

$$\frac{\phi_{cladding}}{\phi_{core}} = \frac{\pi}{8\tan^{-1}\left(\dfrac{2(n-1)f - \sqrt{4(n-1)^2 f^2 - R^2}}{R}\right)} - \frac{1}{2} \tag{7.1}$$

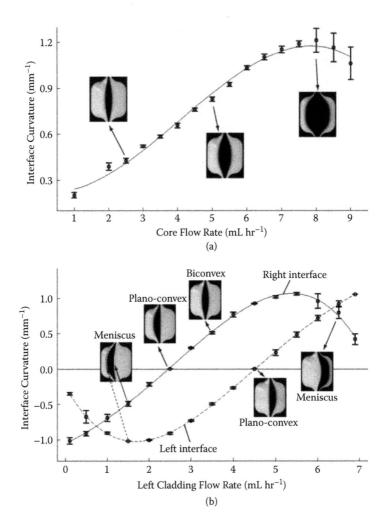

FIGURE 7.10

(a) Curvature ($R_{curvature}$) values of liquid core and liquid cladding lens as function of rates of flow of core stream. The flow rates for the left and right claddings are kept the same. The total flow rate of the core and both claddings is fixed constant at 10 mL/hr. Insets show fluorescence images of lens in expansion chamber at various flow rates. (b) Curvature of left (dashed line) and right (solid line) interfaces of lens as function of flow rate of left cladding. The sign of curvature is positive (negative) if the center of curvature lies to the left (right) of the interface. The flow rate of the core is fixed at 3 mL/hr. The sum of flow rates of the left and right claddings is fixed at 7 mL/hr. Insets show fluorescence images of lens in expansion chamber at indicated flow rates. Error bars are standard deviations of curvatures measured at identical settings of rates of flow in different experimental runs. (*Source:* Tang, S.K.Y., C.A. Stan, and G.M. Whitesides. 2008. *Lab on a Chip,* 8(3), 395–401. With permission.)

Figure 7.13 shows the relationship between the focal length of the lens and the flow rate and refractive index. The focal length was found almost linearly proportional to the flow rate ratio. At the same refractive index, the higher

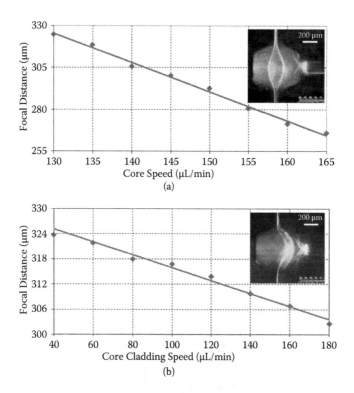

FIGURE 7.11
Focusing of a laser beam using a liquid core and liquid cladding lens (a) with a biconvex lens and (b) with a concave convex lens. (*Source:* Seow, Y.C., A.Q. Liu, L.K. Chin et al. 2008. *Applied Physics Letters*, 93(8), 0841013. With permission.)

the flow rate ratio between the cladding and the core streams, the longer the focal length. At the same flow rate ratio, the higher the refractive index, the shorter the focal length [16].

7.3.4 Optofluidic Microlens Based on Combined Effects of Hydrodynamics and Electro-Osmosis

Li et al. presented a similar design of a circular chamber for lens curvature, adjusted by the combined effects of pressure and electro-osmosis [17]. Figure 7.14 shows the concept. Two cladding streams (fluid 1 and fluid 3) were electrically conducting with high electro-osmotic mobility; the core stream (core fluid 2) was non-conducting with low electro-osmotic mobility.

At a given pressure gradient and voltage applied along the conducting cladding fluids, electro-osmotic forces controlled the curvature of the interfaces (1–2 and 3–2) between the conducting cladding fluids and the core fluid. The curvatures of the interfaces depended on the directions and magnitudes of the applied voltages [17].

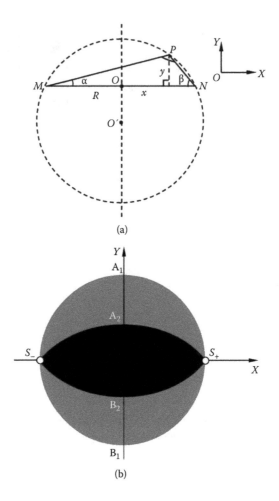

FIGURE 7.12
Mathematical model of optofluidic lens with circular chamber. (*Source:* Song, C., N.T. Nguyen, S.H. Tan et al. 2009. *Lab on a Chip*, 9(9), 1178–1184. With permission.)

Li's group [17] used aqueous sodium chloride (NaCl) solution (7×10^{-4} M) as the cladding liquid. Figure 7.15 demonstrates the interface shapes under different voltages. The measured interface shapes agreed with the calculated curves, confirming that the interface shapes in the circular chamber were optically smooth arcs. The results indicate that the combined effects of hydrodynamics and electro-osmosis may be used to control and tune optofluidic lenses.

7.3.5 Hydrodynamically Adjustable Three-Dimensional Optofluidic Microlenses

Rosenauer et al. demonstrated another fluid manipulation approach (altering flow rates) to form an adjustable three dimensional (3D) optical lens with

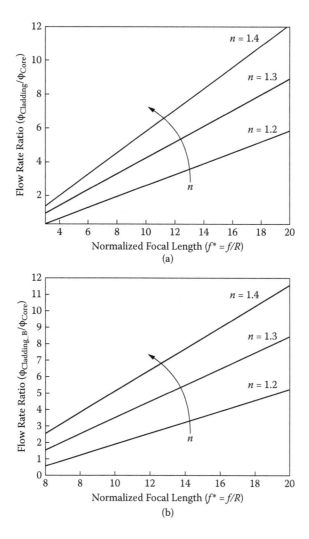

FIGURE 7.13

Relationships between flow rate ratio and focal length of liquid core and liquid cladding lens. The dimension of the focal length was normalized by the radius of the circular chamber. (a) Symmetrical double convex lens. (b) Planoconvex lens. (*Source:* Song, C., N.T. Nguyen, S.H. Tan et al. 2009. *Lab on a Chip*, 9(9), 1178–1184. With permission.)

a single layer, planar microfluidic device [18]. Figure 7.16 depicts the schematics for the formation of a biconic optofluidic lens using computational fluid dynamic (CFD) simulation.

The analyzed fluidic device design utilized three inlets for the lens body and cladding flows and one outlet. For each lens component, there were two more 90-degree curved channels. By combining the single arch channels, a transversal multiconvex lens was generated. The lens body and cladding fluid

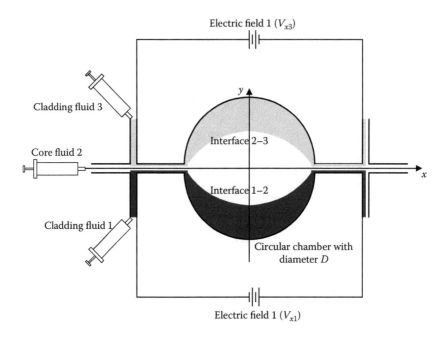

FIGURE 7.14
Optofluidic lens under combined effects of hydrodynamics and electro-osmosis. (*Source:* Li, H., T. Wong, and N.T. Nguyen. 2011. *Microfluidics and Nanofluidics*, 10(5), 1033–1043. With permission.)

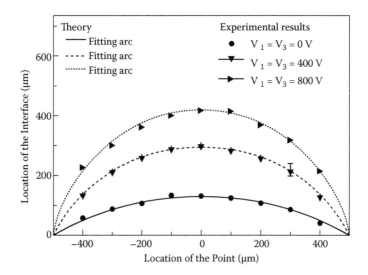

FIGURE 7.15
Shape of interface of optofluidic lens under different applied voltages. (*Source:* Li, H., T. Wong, and N.T. Nguyen. 2011. *Microfluidics and Nanofluidics*, 10(5), 1033–1043. With permission.)

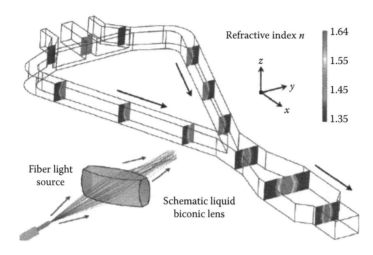

FIGURE 7.16
Microfluidic channel design for 3D optofluidic lens. Cross sections show shaping of lens and refractive index step defined by choice of core and cladding fluids. Light path and lens shape are depicted by ray tracing simulation (Zemax EE). (*Source:* Rosenauer, M. and M.J. Vellekoop. 2009. *Lab on a Chip*, 9(8), 1040–1042. With permission.)

streams were then driven through a lateral expansion region where the flow was widened in the center to produce the 3D character of the optofluidic lens.

By altering the flow rates in absolute values with a constant core-to-cladding ratio, Rosenauer's group [18] could reconfigure the transversal lens independently. The focal plane was adjustable in the y direction as defined in Figure 7.16. Figure 7.17 compares the simulation results and micrographs of the core liquid (stained to show its shape). It shows stable lens symmetry even at high flow rates [18].

7.3.6 Air–Liquid Interfacial Microlens Controlled by Active Pressure

Shi et al. demonstrated another optofluidic microlens utilizing active pressure control of an air–liquid interface [19]. The working mechanism of this microlens is shown in Figure 7.18. DI water was introduced into a straight microchannel. The microchannel made a T-junction with an air reservoir. As the water flowed past the T-junction, air was trapped in the reservoir and a movable air–water interface formed at the T-junction, with the air bending into the water due to the air–liquid contact angle and the hydrophobic–hydrophilic interaction between the surface and the liquid. The air–water interface acted as a divergent lens.

To focus light, a static convergent polymer lens was coupled with the air–water lens. This polymer lens was made of PDMS and bent into an air gap. The focal length of the combination of the two lenses was tuned by adjusting the flow rate of the DI water in the straight microchannel. When this

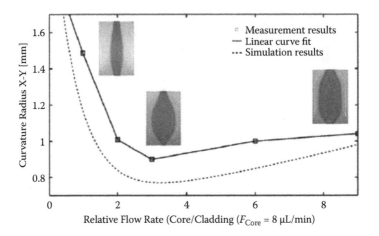

FIGURE 7.17

Radius of curvature of 3D optofluidic lens with respect to the flow rate ratio between core and cladding liquids. Core flow rate is fixed to reshape the lens in the X–Y plane defined in Figure 7.16. The transversal lens is not altered as a function of relative flow rate. Optical micrograph insets show core liquid. Ray tracing simulations of fluidic data with single focal length for both lens directions are presented. (*Source:* Rosenauer, M. and M.J. Vellekoop. 2009. *Lab on a Chip*, 9(8), 1040–1042. With permission.)

flow rate was increased, the air–water interface experienced an increased pressure, resulting in increases in its radius of curvature and in the distance between the interface and the PDMS lens (Figure 7.18b to e).

An increase in the radius of curvature of a divergent air–water interface caused less divergence of the incident light. However, when the interface moved away from the PDMS lens, it generated more divergence in the light reaching the PDMS lens. Thus, the increase in the radius of curvature and the simultaneous movement of the interface exerted counteracting effects on the amount of divergence in the light incident on the PDMS lens.

Nonetheless, the radius of curvature of the interface had a much stronger effect on light divergence than the movement of the interface. Overall, an increased flow rate of the DI water resulted in a shorter image distance. Figure 7.19 shows the experimental results for this lens [19]. Note that increased flow rate of DI water increased the focusing power (or shorter focal length) of the lens combination.

7.4 Tunable Liquid Gradient Refractive Index Lens

All the in-plane tunable microlenses discussed to this point are classical refractive types. Mao et al. reported a tunable microlens configuration based on a liquid gradient refractive index (L-GRIN) lens mechanism [20].

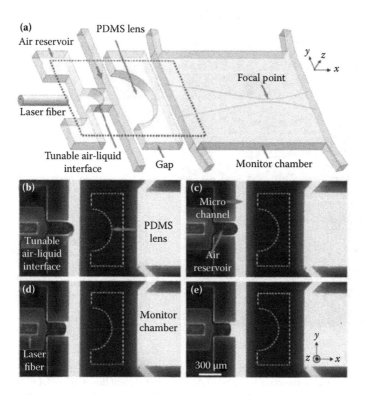

FIGURE 7.18
(a) Schematic of an optofluidic tunable lens. (b) through (e) Optical images of device [region enclosed by dotted square in (a)] taken at different flow rates of DI water in microchannel (*Source:* Shi, J., Z. Stratton, S.C. Lin et al. 2010. *Microfluidics and Nanofluidics*, 9(2), 313–318. With permission.)

A GRIN lens often has a transversely (perpendicular to the optical axis) variable refractive index and a flat lens structure, in contrast to the curved surface of a classical refractive lens. As shown in Figure 7.20A, rather than abruptly changing its direction at the surfaces of classical refractive lenses, light travelling along the optical axis within a GRIN lens is gradually bent toward the optical axis and brought to a focusing point.

More importantly, with precise microfluidic manipulation, one can change both the focal distance of the L-GRIN lens by adjusting the refractive index contrast (difference between maximum and minimum in the gradient) and also the direction of output light by shifting the optical axis [20].

Figure 7.20B1 illustrates the schematics of the L-GRIN lens design. The device includes four inlets and two outlet branches (two branches share a common outlet to maintain the equal back pressure at both sides). Calcium chloride ($CaCl_2$) solution (3.5 M, n ~1.41) and H_2O (n ~1.33) were injected into the device. The refractive index of the mixture was linearly dependent on the $CaCl_2$ concentration.

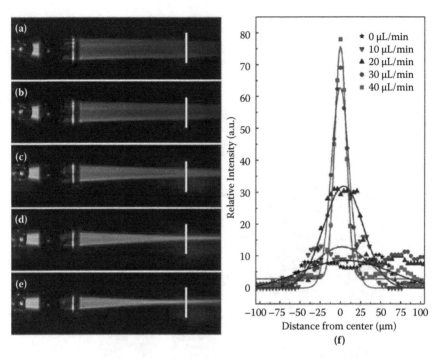

FIGURE 7.19
Ray tracing experiments to characterize variable focal length of and light intensity after combination of air–water-interface lens and PDMS lens at different flow rates of DI water. (a) 0 μL/min. (b) 10 μL/min. (c) 20 μL/min. (d) 30 μL/min. (e) 40 μL/min. (f) Intensity plots along white lines depicted in (a) through (e). Solid curves are fittings to experimental data. (*Source:* Shi, J., Z. Stratton, S.C. Lin et al. 2010. *Microfluidics and Nanofluidics*, 9(2), 313–318. With permission.)

Figure 7.20B2 depicts a typical flow pattern. On each side, two fluids of different refractive indices merge to form co-injected laminar flows and establish a $CaCl_2$ concentration distribution. The convergence of co-injected streams from both sides eventually leads to a complete hyperbolic secant-like refractive index distribution in the main channel. The refractive index profile within the main channel can be adjusted readily by changing the flow rates from different inlets [20].

Figure 7.20C depicts the two operation modes of the L-GRIN lens: the transition mode and the swing mode [20]. The flow rates of $CaCl_2$ solutions remained unchanged in both modes. In the translation mode, the refractive index contrast could be adjusted by symmetrically changing the flow rates of H_2O from both sides to realize different focal lengths such as no focusing (Figure 7.20C1), a large focal length (Figure 7.20C2), and a short focal length (Figure 7.20C3). In the swing mode, the direction of the output light could be adjusted in the device plane by shifting the optical axis of the L-GRIN lens with asymmetrical adjustment of H_2O flow rates from each side (Figure 7.20C3 to C5). The two modes may be operated independently or in combination [20].

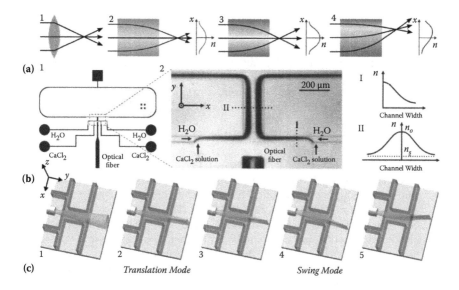

FIGURE 7.20
Principle and design of optofluidic L-GRIN lens. (a) Comparison of classical refractive lens (a1) and GRIN lens (a2). Change in refractive index contrast in GRIN lens can change focal distance (a2 and a3). Shift of optical axis can cause change in output light direction (a4). (b) L-GRIN lens design (b1), microscopic image of L-GRIN lens in operation (b2, left) and expected refractive index distribution at locations I and II inside the lens (b2, right). High optical contrast areas (dark streaks) were observed near fluidic boundaries (b2, left), suggesting significant variation of refractive index from $CaCl_2$ diffusion. (c) Two operation modes of L-GRIN lens: translation mode with variable focal length including no focusing (c1), large focal distance (c2), small focal distance (c3), and swing mode with variable output light direction (c3 through c5). (*Source:* Mao, X.L., S.C.S. Lin, M.I. Lapsley et al. 2009. *Lab on a Chip*, 9(14), 2050–2058. With permission.)

References

1. H. Takiguchi, T. Odake, M. Ozaki, T. Umemura, and K.I. Tsunoda, "Liquid/liquid optical waveguides using sheath flow as a new tool for liquid/liquid interfacial measurements," *Applied Spectroscopy*, vol. 57, pp. 1039–1041, 2003.
2. D. B. Wolfe, R. S. Conroy, P. Garstecki, B. T. Mayers, M. A. Fischbach, K. E. Paul, M. Prentiss, and G. M. Whitesides, "Dynamic control of liquid-core/liquid-cladding optical waveguides," *Proceedings of the National Academy of Sciences of the United States of America*, vol. 101, pp. 12434–12438, 2004.
3. J. Y. Kim, K. H. Jeong, and L. P. Lee, "Artificial ommatidia by self-aligned microlenses and waveguides," *Optics Letters*, vol. 30, pp. 5–7, 2005.
4. D. A. Chang-Yen, R. K. Eich, and B. K. Gale, "A monolithic PDMS waveguide system fabricated using soft-lithography techniques," *Journal of Lightwave Technology*, vol. 23, pp. 2088–2093, 2005.
5. R. S. Conroy, B. T. Mayers, D. V. Vezenov, D. B. Wolfe, M. G. Prentiss, and G. M. Whitesides, "Optical waveguiding in suspensions of dielectric particles," *Applied Optics*, vol. 44, pp. 7853–7857, 2005.

6. K. H. Jeong, J. Y. Kim, and L.P. Lee, "Biologically inspired artificial compound eyes," *Science*, vol. 312, pp. 557–561, 2006.
7. S. K. Y. Tang, B. T. Mayers, D. V. Vezenov, and G. M. Whitesides, "Optical waveguiding using thermal gradients across homogeneous liquids in microfluidic channels," *Applied Physics Letters*, vol. 88, p. 061112, 2006.
8. D. V. Vezenov, B. T. Mayers, D. B. Wolfe, and G.M. Whitesides, "Integrated fluorescent light source for optofluidic applications," *Applied Physics Letters*, vol. 86, p. 041104, 2005.
9. S. Camou, H. Fujita, and T. Fujii, "PDMS 2D optical lens integrated with microfluidic channels: principle and characterization," *Lab on a Chip*, vol. 3, pp. 40–45, 2003.
10. S.-K. Hsiung, C.-H. Lin, G.-B. Lee, "A microfabricated capillary electrophoresis chip with multiple buried optical fibers and microfocusing lens for multiwavelength detection," *Electrophoresis*, vol. 26, pp. 1122–1129, 2005.
11. G.-B. Lee, S.-H. Chen, G.-R. Huang, W.-C. Sung, and Y.-H. Lin, "Microfabricated plastic chips by hot embossing methods and their applications for DNA separation and detection," *Sensors and Actuators B: Chemical*, vol. 75, pp. 142–148, 2001.
12. L. Dong, and H. Jiang, "Tunable and movable liquid microlens *in situ* fabricated within microfluidic channels," *Applied Physics Letters*, vol. 91, p. 041109, 2007.
13. X. Mao, J. R. Waldeisen, B. K. Juluri and T. J. Huang, "Hydrodynamically tunable optofluidic cylindrical microlens," *Lab on a Chip*, vol. 7, pp. 1303–1308, 2007.
14. S. K. Y. Tang, C. A. Stan, and G. M. Whitesides, "Dynamically reconfigurable liquid-core liquid-cladding lens in a microfluidic channel," *Lab on a Chip*, vol. 8, pp. 395–401, 2008.
15. Y. C. Seow, A. Q. Liu, L. K. Chin, X. C. Li, H. J. Huang, T. H. Cheng, and X. Q. Zhou, "Different curvatures of tunable liquid microlens via the control of laminar flow rate," *Applied Physics Letters*, vol. 93, p. 084101, 2008.
16. C. Song, N. T. Nguyen, S. H. Tan, and A. K. Asundi, "Modelling and optimization of micro optofluidic lenses," *Lab on a Chip*, vol. 9, pp. 1178–1184, 2009.
17. H. Li, T. Wong, and N.-T. Nguyen, "A tunable optofluidic lens based on combined effect of hydrodynamics and electroosmosis," *Microfluidics and Nanofluidics*, vol. 10, pp. 1033–1043, 2011.
18. M. Rosenauer and M. J. Vellekoop, "3D fluidic lens shaping—a multiconvex hydrodynamically adjustable optofluidic microlens," *Lab on a Chip*, vol. 9, pp. 1040–1042, 2009.
19. J. Shi, Z. Stratton, S.-C. Lin, H. Huang, and T. J. Huang, "Tunable optofluidic microlens through active pressure control of an air–liquid interface," *Microfluidics and Nanofluidics*, vol. 9, pp. 313–318, 2010.
20. X. Mao, S.-C. Lin, M. I. Lapsley, J. Shi, B. K. Juluri, and T. J. Huang, "Tunable Liquid Gradient Refractive Index (L-GRIN) lens with two degrees of freedom," *Lab on a Chip*, vol. 9, pp. 2050–2058, 2009.

8

Looking into the Future

In this chapter, we first summarize our previous discussions and recap the importance of microlenses. Then we will attempt to shed light onto the future directions related to the research and development of microlenses, specifically the challenges such as packaging of the complete system, effects of gravity, evaporation of liquids, aberrations, and integration with other optical components.

8.1 Commercialization of Microlenses

In the previous chapters, we discussed the importance of microlenses in communications, imaging, lithography, displays, and sensors. Although microlenses have undergone rapid development in recent years, most studies of microlenses, especially tunable types, are still preliminary, and their market applications remain limited. The most common applications are solid microlens arrays, as discussed in Chapter 4. Microlenses with different characteristics such as size, focal length, and fill factor are commercially available.

Currently, certain issues limit the commercialization of microlenses. Addressing these issues will shape the future development of these devices.

8.2 Future Work

8.2.1 Packaging

Packaging is a key factor for microlenses as it is for electronics and photo-electronics. Unfortunately, packaging for microlenses presents even bigger challenges than those usually faced by electronic chips because microlenses involve both electrical interfaces and optical interfaces and material compatibilities with environments [1]. However, some initial success of packaging has already been achieved.

TABLE 8.1

Design Parameters of Lens Module
Shown in Figure 5.30

Parameter	Value
VGA CMOS sensor	640 × 480 pixels
Pixel size	5.0 ×5.0 μm²
Field of view	60 degrees
F number	2.5
Height	5.5 mm
Entrance pupil diameter	1.43 mm
Focal range	2 cm → ∞

Source: Hendriks, B.H.W. et al. 2005. *Optical Review,* 12, 255–259. With permission.

Kuiper et al. described an optical module design for an electrowetting-based microlens [2], as shown in Figure 5.30. Table 8.1 lists the design parameters of the lens module [3].

Figure 6.34 demonstrates the schematics of a prototype endoscope with a liquid tunable-focus microlens integrated at the distal end. All optical fibers and the fiber bundles were bound together by adapters and attached to the back side of the microlens. Figure 8.1 depicts a prototype endoscope with a tunable-focus microlens integrated at its end [4]. One optical fiber bundle (2 mm diameter, 17,000 pixels) was used for acquiring images in the prototype endoscope. Two sets of adapters were used. One was designed to connect the liquid guide from the light sources to the actuation optical fibers, and the other was for connecting the actuation optical fibers and the image acquisition fiber bundle to the back side of the glass slide with the microlens. All adapters were made of aluminum alloy or polycarbonate and fixed by set screws. The transmission efficiency of IR light from a source to the hydrogel microstructures was measured to be less than 3%.

The examples in Figure 6.34 and Figure 8.1 demonstrate the feasibility of packaging liquid-based microlenses into applicable systems. Nonetheless, further investigation of packaging is needed.

8.2.2 Effects of Inertia

Forces in microsystems decrease as size is scaled down. Different forces scale down into the microdomain differently, depending on the dimensions [5]. The forces were found to scale in one of four ways following the applicable physical laws. If the scale size is decreased by a factor of 10, the forces may decrease according to the various physical laws by 10, 100, 1,000, or 10,000 times, respectively. For example, at the same downsizing rate, surface tension decreases only by a factor of 10, while gravity decreases by a factor of 1,000.

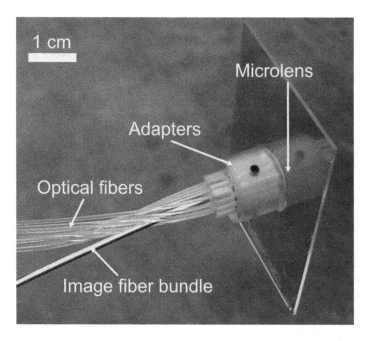

FIGURE 8.1

Photograph of prototype endoscope with tunable-focus liquid microlens at distal end. An image acquisition fiber bundle and 12 optical fibers are bound together and attached to the back side of the microlens by adapters. (*Source:* Zeng, X. et al. 2011. *Journal of Microelectromechanical Systems,* 20, 583–593. With permission.)

The scaling of surface tension is advantageous and may dominate gravity when the feature size of the system is smaller than a specific value. Normally such feature value for most liquids is close to 1 mm^3 (10^{-3} mL) [6,7], as shown by an example in Chapter 2. Therefore, for most microlenses whose sizes are smaller than or near 1 mm, the effect of gravity may be neglected.

However, in some cases, multiple liquids may be used in a single microlens, and their differences in gravity could raise some issues. To reduce the effect of inertia from the different densities of the liquids, Kuiper and Hendriks matched the liquid density in an electrowetting-based lens by using a salt solution [2]. Dissolution of 20% lithium chloride led to a density (ρ) of 1.12 kg/m^3. They used a mixture of phenylmethylsiloxanes as the insulating liquid because of their high refractive indices and good electrowetting properties. A few percent of dense carbon tetrabromide (ρ = 2.96 kg/m^3) was dissolved in the insulating liquid to match the density of the salt solution.

Theisen et al. presented a dynamic model to describe the oscillations resulting from the competition between liquid inertia and capillarity [8]. They demonstrated that gravity was weaker than surface tension but exerted a non-negligible effect. It was found that gravity could distort the shapes of spherical droplets. Measures had to be taken to counter the distortion, such as using localized pressure bias onto droplets to offset the changes in shapes. Such methods shed light onto how to counter the gravity issue and warrant further exploration.

8.2.3 Evaporation of Liquids

Most microlenses use water or water-based solutions; thus prevention of water evaporation in these systems presents another challenge for packaging. In the microlenses reported by Jiang et al., as discussed in Section 6.5, water was actuated by hydrogels and was covered by oil for the prevention of evaporation [9–11]. In the electrowetting-based lenses, oil was used for the same purpose [2,12].

For some mechanically-driven microlenses (Section 6.3) and electrostatic microlenses (Section 5.3), the liquid chambers were covered by a layer of thin membrane and thus water was prevented from leaving the chamber [13–16]. More careful study of sealing liquid containers to prevent evaporation should be conducted in the future.

8.2.4 Aberrations

Aberration of microlenses is a key parameter for optical components. We discussed aberration in Chapter 2. For most microlens systems, the lenses are manipulated and controlled by surface tension or their formation procedure involves surface tension. As a result, the shapes of these microlenses are mainly defined by surface tension.

Spherical aberrations of single microlenses have been studied more extensively [17,18], although other types of aberrations were measured too. Different from traditional macro-sized optical lenses, the curvatures of microlenses cannot be changed locally. Hence, one approach suggested to reduce aberrations is to assemble multiple lenses along the optical axis [19–21]. However, this and other approaches proposed to decrease aberrations must be investigated further.

8.2.5 Integration into Systems

Most microlenses reported to date are single component types. Light is transmitted into the microlenses through extra lenses or optical fibers. With increasing emphasis on system integration, it has become vital for microlenses to be implemented seamlessly into complete systems.

As noted earlier, Zeng et al. showed an endoscope integrated with a tunable microlens at the tip of an optical fiber [4]. Weber et al. integrated two lenses and one aperture onto a micro-optical bench for optical coherence tomography [21]. Optofluidic systems discussed in Chapter 7 are also good examples of exploring the integration of optical components in a system. For instance, tunable microlenses and optical fibers were formed in microchannels and used to observe objects in a closed channel [22–26]. With further research and investigation, the issue of integration of microlenses should be resolved.

References

1. A. P. Malshe, C. O'Neal, S. B. Singh, W. D. Brown, W. P. Eaton, and W. M. Miller, "Chanllenges in the Packaging of MEMS," *International Journal of Microcircuits and Electronic Packaging*, vol. 22, pp. 233–241, 1999.
2. S. Kuiper and B. H. W. Hendriks, "Variable-focus liquid lens for miniature cameras," *Applied Physics Letters*, vol. 85, pp. 1128–1130, 2004.
3. B. H. W. Hendriks, S. Kuiper, M. A. J. VAN As, C. A. Renders, and T. W. Tukker, "Electrowetting-based variable-focus lens for miniature systems," *Optical review*, vol. 12, pp. 255–259, 2005.
4. X. Zeng, C. T. Smith, J. C. Gould, C. P. Heise, and H. Jiang, "Fiber endoscopes utilizing liquid tunable-focus microlenses actuated through infrared light," *Journal of Microelectromechanical Systems*, vol. 20, pp. 583–593, 2011.
5. W. S. N. Trimmer, "Microrobots and micromechanical systems," *Sensors and Actuators*, vol. 19, pp. 267–287, 1989.
6. X. Zeng, *Manipulation and Simulation of Droplets Based on Electrowetting on Dielectrics*. Institute of Microelectronics M.S. Thesis, Beijing: Tsinghua University, 2005.
7. C.-J., Kim, "MEMS devices based on the use of surface tension," *Proc. Int. Semiconductor Device Research Symposium (ISDRS'99)*, Charlottesville, VA, pp. 481–484, Dec 1999.
8. E. A. Theisen, M. J. Vogel, C. A. López, A. H. Hirsa, and P. H. Steen, "Capillary dynamics of coupled spherical-cap droplets," *Journal of Fluid Mechanics*, vol. 580, pp. 495–505, 2007.
9. L. Dong, A. K. Agarwal, D. J. Beebe and H. Jiang, "Adaptive liquid microlenses activated by stimuli-responsive hydrogels," *Nature*, vol. 442, pp. 551–554, 2006.
10. X. Zeng and H. Jiang, "Tunable liquid microlens actuated by infrared light-responsive hydrogel," *Applied Physics Letters*, vol. 93, pp. 151101, 2008.
11. L. Dong, A. K. Agarwal, D. J. Beebe, and H. Jiang, "Variable-focus liquid microlenses and microlens arrays actuated by thermoresponsive hydrogels," *Advanced Materials*, vol. 19, pp. 401–405, 2007.
12. C. X. Liu, J. Park, and J. W. Choi, "A planar lens based on the electrowetting of two immiscible liquids," *Journal of Micromechanics and Microengineering*, vol. 18, p. 035023, 2008.
13. H. Ren and S. T. Wu, "Variable-focus liquid lens by changing aperture," *Applied Physics Letters*, vol. 86, p. 211107, 2005.
14. H. W. Ren, and S. T. Wu, "Variable-focus liquid lens," *Optics Express*, vol. 15, pp. 5931–5936, 2007.
15. K. H. Jeong, G. L. Liu, N. Chronis, and L. P. Lee, "Tunable microdoublet lens array," *Optics Express*, vol. 12, pp. 2494–2500, 2004.

16. N. Binh-Khiem, K. Matsumoto, and I. Shimoyama, "Polymer thin film deposited on liquid for varifocal encapsulated liquid lenses," *Applied Physics Letters*, vol. 93, p. 124101, 2008.

17. X. Zeng and H. Jiang, "Polydimethylsiloxane microlens arrays fabricated through liquid-phase photopolymerization and molding," *Journal of Microelectromechanical Systems*, vol. 17, pp. 1210–1217, 2008.

18. N. Chronis, G. L. Liu, K.-H. Jeong, and L. P. Lee, "Tunable liquid-filled microlens array integrated with microfluidic network," *Optics Express*, vol. 11, pp. 2370–2378, 2003.

19. D. Y. Zhang, N. Justis, and Y. H. Lo, "Fluidic adaptive lens of transformable lens type," *Applied Physics Letters*, vol. 84, pp. 4194–4196, 2004.

20. L. Pang, U. Levy, K. Campbell, A. Groisman, and Y. Fainman, "Set of two orthogonal adaptive cylindrical lenses in a monolith elastomer device," *Optics Express*, vol. 13, pp. 9003–9013, 2005.

21. N. Weber, H. Zappe, and A. Seifert, "A tunable optofluidic silicon optical bench," *Journal of Microelectromechanical Systems*, vol. 21, pp. 1357–1364, 2012.

22. L. Dong and H. Jiang, "Tunable and movable liquid microlens in situ fabricated within microfluidic channels," *Applied Physics Letters*, vol. 91, p. 041109, 2007.

23. X. Mao, J. R. Waldeisen, B. K. Juluri and T. J. Huang, "Hydrodynamically tunable optofluidic cylindrical microlens," *Lab on a Chip*, vol. 7, pp. 1303–1308, 2007.

24. S. K. Y. Tang, C. A. Stan, and G. M. Whitesides, "Dynamically reconfigurable liquid-core liquid-cladding lens in a microfluidic channel," *Lab on a Chip*, vol. 8, pp. 395–401, 2008.

25. J. Shi, Z. Stratton, S.-C. Lin, H. Huang, and T. J. Huang, "Tunable optofluidic microlens through active pressure control of an air–liquid interface," *Microfluidics and Nanofluidics*, vol. 9, pp. 313–318, 2010.

26. X. Mao, S.-C. Lin, M. I. Lapsley, J. Shi, B. K. Juluri, and T. J. Huang, "Tunable Liquid Gradient Refractive Index (L-GRIN) lens with two degrees of freedom," *Lab on a Chip*, vol. 9, pp. 2050–2058, 2009.

Index

2-(Dimethylamino)ethyl methacrylate (DMAEMA), 166
2-Hydroxyethyl methacrylate (HEMA), 166

A

Abbe number, 14
Aberration, 13, 17, 203
    Astigmatism, 20–21, 130
    Chromatic aberration, 21–22
    Coma, 18–20
    Distortion, 21–22, 39, 78, 148, 175, 206
    Spherical aberration , 17–18, 26–30, 148, 206
Absorption, 14
Acrylic acid (AA), 164, 166
Additive process, 44
Adhesive force, 31–33, 39
Air-liquid interface (liquid-air interface), 6, 11, 23, 98, 185, 197
Airy disk, 25
Airy pattern, 25
Angular resolution, 25
Annealing, 43, 55, 61, 80
    Rapid thermal annealing (RTA), 55
Anomalous dispersion, 13
ANSYS, 147, 150
Aperture, 3, 6, 24, 26, 29, 38, 68, 94, 97–99, 117, 148, 155–159, 165–167, 188, 190, 206
Atomic force microscopy (AFM), 75
Axial chromatic aberration, 22

B

Beam shaping, 7
Biconvex, 148, 154–155, 190–191
Bifocal (bifocal lens), 148, 150–151
Birefringence (birefringent), 108–109, 112, 117
Blank exposure, 66
Bond number , 127, 140

Bonding, xiii, 43, 54, 59–60, 69, 153–155, 182
    Anodic bonding, 60, 145
    Fusion bonding, 60

C

Calix[4]hydroquinone (CHQ), 76–77
Capillary action, 32, 110
Centrifugal effect, 188
Centrifugal force, 188
Charge-coupled device (CCD), 7, 23, 29, 86, 96, 113, 116–118, 150, 169–171, 173
Chemical mechanical polishing (CMP), 61, 94
Chemical treatment, 35
Chemical vapor deposition (CVD), 48, 58–59
    Low pressure chemical vapor deposition (LPCVD) , 46, 48, 58–59
    Ultrahigh vacuum chemical vapor deposition (UHVCVD) , 72
    Plasma enhanced chemical vapor deposition (PECVD), 48
Cladding stream, 190, 193
Clausius-Mossotti (CM) factor, 82
Clean room, xiv, 44
CMOS, 137, 204
Cohesive force, 30–33
Collimated beam (collimated light, collimated illumination), 5, 29, 88, 117, 148, 151
Comatic flare, 20
Compound eye, xiii, 1, 4, 9, 174
Compound semiconductor, 45
Computational fluid dynamics (CFD), 195
Conjugate distance , 15
Conjugate plane, 15
Conjugate point, 15
Contact aligner, 49